A PRIMER ON STRING THEORY

Since its conception in the 1960s, string theory has been hailed as one of the most promising routes we have to unify quantum mechanics and general relativity. This book provides a concise introduction to string theory, explaining central concepts and mathematical tools and covering recent developments in physics, including compactifications and gauge/string dualities. With string theory being a multidisciplinary field interfacing with high-energy physics, mathematics, and quantum field theory, this book is ideal for both students with no previous knowledge of the field and scholars from other disciplines who are looking for an introduction to basic concepts.

VOLKER SCHOMERUS holds a joint position as Senior Researcher in the DESY Theory Group and Professor of Mathematical Physics at Universität Hamburg. His work focuses on string theory at the interface of high-energy physics, mathematics, and statistical physics, for which he has received several distinctions, including the Guy-Lussac-Humboldt award in 2010. He is also co-author of the monograph *Boundary Conformal Field Theory and the World-Sheet Approach to D-Branes* (Cambridge, 2013).

A PRIMER ON STRING THEORY

VOLKER SCHOMERUS

Deutsches Elektronen-Synchrotron (DESY) Hamburg

CAMBRIDGE
UNIVERSITY PRESS

CAMBRIDGE
UNIVERSITY PRESS

University Printing House, Cambridge CB2 8BS, United Kingdom

One Liberty Plaza, 20th Floor, New York, NY 10006, USA

477 Williamstown Road, Port Melbourne, VIC 3207, Australia

314-321, 3rd Floor, Plot 3, Splendor Forum, Jasola District Centre, New Delhi-110025, India

79 Anson Road, #06-04/06, Singapore 079906

Cambridge University Press is part of the University of Cambridge.

It furthers the University's mission by disseminating knowledge in the pursuit of education, learning and research at the highest international levels of excellence.

www.cambridge.org
Information on this title: www.cambridge.org/9781107160019
DOI: 10.1017/9781316672631

First published 2017

A catalogue record for this publication is available from the British Library

ISBN 978-1-107-16001-9 Hardback
ISBN 978-1-316-61283-5 Paperback

Contents

Preface

About half a century has passed since string theory was born, initially to explain the observed Regge trajectories of hadronic resonances. During these past decades, it has gone through several phases of dismissal and new appreciation, it has become a promising candidate for a quantum theory of gravity, and occasionally it has been popularized as a theory of everything. What most of these periods have in common are profound and fruitful interactions with rather different areas of theoretical physics, e.g., with quantum field theory, high-energy physics, gravity, mathematics, and even statistical and condensed matter physics. Indeed, there is a long list of string-inspired developments, a few of which we will have a chance to touch upon below. Among them are, e.g., new unified extensions of the standard model that are being tested in collider experiments, insights into black holes and their entropy, contributions to knot invariants or the theory of modular forms, and our understanding in particular of 2-dimensional critical systems, or the impressive recent progress in accessing both perturbative as well as non-perturbative effects in supersymmetric gauge theories. This interdisciplinary nature of string theory is its real strength, and it can serve as a good motivation to enter or explore the field.

While this textbook addresses primarily masters' or early PhD students, it may also be useful for scholars from neighboring fields who are looking for a short exposition of basic string theory. Compared to most other textbooks in the field, the scope here is fairly modest. Rather than attempting to cover string theory from its roots to all the fascinating modern developments, the intention is merely to provide a set of concepts and tools that are common to a wide range of recent research directions. By now many books have been written on string theory, which include various advanced and topical directions, such as [27, 28, 55, 56, 6, 39, 74, 41, 1] to list just a few that can be turned to for further reading. In addition, let me also mention a few shorter introductions such as [65, 68, 2]. Finally, a beautiful book by Barton Zwiebach [80] covers much less material but addresses undergraduates.

Some students might find it useful to consult this text, in particular to fill in occasional gaps in their undergraduate education.

The restrictions imposed on the choice of topics make it possible to cover the material of these Chapters in a one or two semester introductory course. Very few prerequisites are assumed, e.g., from quantum field theory, general relativity, or supersymmetry, so that master's students should have no problem following the exposition. The book consists of two parts. Part I describes strings in flat backgrounds. It introduces most of the basic constructions and results in string theory. This part of the material can be covered in a 10 to 12 lecture crash course, including Chapters 1–4, 6–8, and 10, along with Chapters 18 and 20 from Part II, for an outlook on modern developments. In this short form, the chapters assume some basic knowledge of supersymmetry. The minimal background is collected in Chapter 9. The remaining chapters from Part I deal with light-cone quantization, D-branes, and heterotic strings. These may be considered as additional material that may be included if time permits.

Part II contains somewhat more advanced topics on strings in curved backgrounds and string compactifications. It addresses three distinct subtopics. The first one is *conformal field theory*. As an example of the free field some very basic strategies and notions are presented in Chapter 13. The concept of modular invariance is the central theme of Chapter 14. It is explored with the example of the so-called orbifold construction. The third chapter on conformal field theory techniques deals with the SU(2) WESS-ZUMINO-NOVIKOV-WITTEN model, a theory that can be used to describe strings on a space of constant curvature. It allows us to illustrate most of the key techniques and concepts of conformal field theory.

The second subtopic treats CALABI-YAU spaces and the associated string *compactifications*. Chapter 16 introduces some background material on geometric concepts starting from complex manifolds and ending with some basic examples of complete intersections of CALABI-YAU spaces such as the famous quintic. This background material is then applied to discuss basic properties of string compactifications on CALABI-YAU spaces. The focus is on a space-time analysis, but a short discussion of the relevant world-sheet models is also provided, thereby interacting with the chapters on conformal field theory.

The book concludes with three chapters on *string dualities*. These possess the character of an overview rather than a detailed exposition. Chapter 18 starts with a brief discussion of T-duality, proceeds to the self-duality of type IIB superstring theory, and finishes with strongly coupled IIA models and some evidence for the existence of M-theory. Chapters 19 and 20 finally deal with the dualities between gauge theory and string theory that appear in the context of the *AdS/CFT correspondence*. These include a short discussion of large N_c limits before stating Maldacena's duality between the maximally (N=4) supersymmetric YANG-MILLS

(SYM) theory in four dimensions and strings in an $AdS_5 \times S^5$ background. Some basic entries of the dictionary between string and gauge theory are also introduced. The final chapter carries Maldacena's duality a bit deeper by providing more detail on N=4 SYM theory and on semiclassical strings in $AdS_5 \times S^5$.

I gave the lectures that make up this book or some of them on various occasions, in particular during the "String Steilkurs" at DESY, as one or two semester courses in Hamburg University, and at the APCTP Winter School. Many of the students who attended these courses have contributed through their questions and comments. I wish to thank all of them and in particular those who spent their time to compile seemingly never ending lists of misprints. I am particularly endebted to Philipp Höffer v. Loewenfeld and Johannes Oberreuter, who initiated these notes and typed the first 10 chapters. Some of the more advanced chapters were written with the assistence of my former students Vladimir Mitev and Maike Tormählen. Finally, I am grateful to Simon Capelin for his continued encouragement to complete the notes.

I dedicate this book to my wife Elena, to our sons Jonathan and Benedict, and to Leon.

1

Historical Introduction and Overview

In this first chapter, I shall review some of the history of string theory, starting with early attempts to apply strings to hadronic physics. Then, I discuss how string theory emerged as a theory of gravitational physics and sketch the picture of modern string theory that was developed mostly during the second half of the 1990s. The discussion is based on images. It is the main goal of this book to introduce all the background that is required to give precise meaning to the cartoon you are about to see (Figures 1.2–1.6). In a first approach, the present chapter just serves for some rough and qualitative orientation.

1.1 String Theory and Strong Interactions

Most insight into the physics of particles comes from the study of scattering amplitudes. To take the simplest case, let us consider an elastic scattering process of two particles. After the scattering event, two particles emerge from the scattering event.

Such a $2 \to 2$ scattering process possesses many symmetries. The amplitude can depend only on kinematic invariants of the process. A widely used choice for these invariants is given by the so-called MANDELSTAM variables s, t, u, which are defined by

$$s = -(p_1 + p_2)^2, \tag{1.1}$$

$$t = -(p_1 + p_3)^2, \tag{1.2}$$

$$u = -(p_1 + p_4)^2. \tag{1.3}$$

Here, p_1 and p_2 denote the momenta of the incoming particles, and $-p_3, -p_4$ are momenta of the outgoing ones, see Figure 1.2. Throughout this book we shall work with a MINKOWSKI signature in which

$$q^2 = -q_0^2 + q_1^2 + \cdots + q_{D-1}^2.$$

1

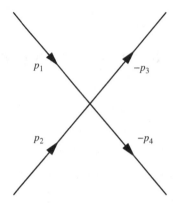

Figure 1.1 Kinematics of an elastic $2 \to 2$-scattering process of particles with momenta p_i.

The sum of the three MANDELSTAM variables is determined by the masses m_i of the particles through $s + t + u = \sum_i m_i^2$. From now on we shall consider amplitudes as functions of s, t.

In the center of mass system, we can reexpress s and t through the energy $E = E_1 = E_2$ of the incoming particles and the center of mass scattering angle θ_s:

$$s = 4E_1^2 , \tag{1.4}$$

$$t = -\frac{s}{2}(1 - \cos \theta_s) . \tag{1.5}$$

In the late 1960s, physicists analyzed in great detail the scattering amplitudes for hadronic particles. What they found in the laboratory was an enormous number of new resonances. The longer the searches were pursued, the higher became the spins J and masses M of the observed particles. In addition, a curious linear relation $J = \alpha_0 + \alpha' M^2$ emerged that could be characterized by the so-called REGGE slope α' and intercept α_0; see, e.g. [37, 16]. Since at high energies s the t-channel exchange of a spin J resonance with mass M_J contributes a term $g_J^2(-s)^J/(t - M_J^2)$ to the total amplitude,[1] the overall contribution of the hadronic resonances takes the form

$$\mathcal{A}(s,t) = \sum_J \quad\sim\quad -\sum_J g_J^2 \frac{(-s)^J}{t - M_J^2} , \tag{1.6}$$

where the sum is taken over all possible intermediate resonances that can be exchanged in the t channel. If there was only a finite number of such terms, i.e., if the sum stopped at some finite value of J, then the high-energy behavior

---

[1] The expression we work with here is valid only for large values of s. It receives lower order corrections for finite s that are well known but not relevant for our discussion.

would be dominated by the resonance with highest spin J, and the whole amplitude would diverge as we send s to infinity. If the sum is infinite, however, the outcome may be very different. As an example, let us consider the exponential function

$$\exp(-x) = 1 - x + \frac{1}{2}x^2 - \frac{1}{3!}x^3 + \ldots.$$

Suppose now that we truncate this sum after the nth term. Then, the finite sum diverges for $x \to \infty$. On the other hand, the entire function $\exp(-x)$ is certainly finite. In fact, it even vanishes in the limit $x \to \infty$. Could something similar happen for hadronic scattering amplitudes, i.e., could there be some analytic expression $\mathcal{A}$ that reproduces the expansion (1.6) for small s but that stays finite at large s? The answer is positive, and one such expression is given by the so-called VENEZIANO amplitude [71]

$$\mathcal{A}_{\text{Ven}}(s, t) = \frac{\Gamma(-\alpha_0 - \alpha' s)\Gamma(-\alpha_0 - \alpha' t)}{\Gamma(-2\alpha_0 - \alpha'(s+t))} . \tag{1.7}$$

From the pole structure and shift properties of Γ functions it is easy to deduce the following expansion of $\mathcal{A}$ at small s:

$$\mathcal{A}_{\text{Ven}}(s, t) \sim -\sum_J \frac{P^J(s)}{\alpha' t - J + \alpha_0} . \tag{1.8}$$

Here, P^J is a polynomial of degree J. Hence, $\mathcal{A}$ does indeed encode the exchange of resonances that lie on a REGGE trajectory $M^2 = (J - \alpha_0)/\alpha'$. In 1970, string theory seemed to provide an exciting perspective on these findings. Namely, it was shown that simple (open) string theories in flat space [47, 50, 64] naturally reproduce the VENEZIANO amplitude with $\alpha_0 = 1$. This success of string theory is not too difficult to understand. String modes in flat space are harmonic oscillators, and it is well known from basic quantum mechanics that these possess a linear spectrum with a distance between the spectral lines that is determined by the tension T_s of the string. If we choose the latter to be $T_s \sim 1/\alpha'$, then we may identify hadronic resonances with vibrational modes of a string (provided we are willing to close an eye to the first resonance with $J = 0$, which is tachyonic because $\alpha_0 = 1$).

Obviously, the formula (1.7) must not be restricted to small center of mass energies. It can also be evaluated, e.g., for fixed angle scattering at large s. Using once more some simple properties of the Γ function one can derive

$$\mathcal{A}_{\text{Ven}}(s, t) \sim f(\theta_s)^{-1-\alpha' s} \quad (\text{large } s) , \tag{1.9}$$

where f is some function of the center of mass scattering angle θ_s whose precise form is not relevant for us. The result shows that fixed angle scattering amplitudes predicted by flat space string theory fall off exponentially with the energy $\sqrt{s}$. Unfortunately for early string theory, this is not at all what is found in experiments, which display much harder high-energy cross sections. The failure of string theory to produce the correct high-energy features of scattering experiments is once more easy to understand: strings are extended objects, and as such they do not interact in a single point but rather in an extended region of space-time. Consequently, their scattering amplitudes are rather soft at high energies (small distances), at least compared to point particles. In this sense, experiments clearly favored a point particle description of strongly interacting physics over fundamental GeV scale strings.

As we all know, a highly successful point particle model for strong interactions, known as Quantumchromodynamics (QCD), was established only a few years later. It belongs to the class of gauge theories that have ruled our description of nature for several decades now. Due to its asymptotic freedom, high-energy QCD is amenable to a perturbative treatment. On the other hand, low-energy (large distance) physics is strongly coupled and therefore remains difficult to address. Even though the problem to understand, e.g., confinement remains unsolved, QCD has at least never made any predictions that could be clearly falsified in a simple laboratory experiment, in contrast to what we have reviewed about early string theory. So, in spite of its intriguing success with hadronic resonances, string theory retreated from the area of strong interactions, and it even disappeared from physics before reemerging as a quantum theory of gravity.

1.2 Closed Strings and Supergravity

Before we can fully appreciate the role of closed string theory as a natural host for gravity, we would like to briefly recall the basic problem one faces when applying perturbation theory to quantize EINSTEIN's theory of gravity. In order to see the issue, let us compare the following two amplitudes:

$$\mathcal{A}_0 = \quad , \qquad \mathcal{A}_1 = \quad . \tag{1.10}$$

The second amplitude comes weighted with NEWTON's gravitational constant G_N. Therefore, by a simple dimensional analysis, we conclude that the dimensionless quotient of $\mathcal{A}_1/\mathcal{A}_0$ must be given by

$$\frac{\mathcal{A}_1}{\mathcal{A}_0} \sim \frac{G_N E^2}{\hbar c^5}, \tag{1.11}$$

i.e., it is proportional to E^2, a behavior that also follows directly from the fact that the gravitational interaction is mediated by a massless particle of spin $J = 2$ $(s^2/t \sim E^2)$. From this formula it is clear that quantum corrections are divergent for large energies. A similar analysis of quantum corrections shows that the divergence becomes worse and worse as we go to higher loop order. This failure of perturbative quantum gravity has led to the conclusion that EINSTEIN's theory is unlikely to be a fundamental theory of gravity. Similarly to FERMI's theory of weak interactions, it should rather be considered as an effective low-energy theory that must be deformed at high energies in order to be consistent with the principles of quantum physics.

As we mentioned before, superstring theory re-emerged in the 1980s after it had been realized that it provided a natural and consistent host for gravitons [29]. In order to be a bit more specific, we shall consider closed strings propagating in some background geometry X. It is widely known that superstrings require X to be 10-dimensional, so that contact with 4-dimensional physics is often made by rolling extra directions up on small circle or through more general compactifications. The study of such compactifications is an extremely active field of research. We will get back to the issue in the second part of this book.

Strings possess infinitely many vibrational modes that we can think of as an infinite tower of massless and massive particles propagating on X. The mass spectrum of the theory is linear, with the separation that is parametrized by the tension $T_s \sim 1/\alpha'$ or, equivalently, by the length $l_s = \sqrt{\alpha'}$ of the string. As strings propagate through X, they can interact by joining and splitting. One simple such process for a one-loop contribution to the $2 \rightarrow 2$ scattering of closed strings is depicted in Figure 1.3. On the left-hand side we have drawn a one-loop diagram in some arbitrary field theory with 3- and 4-point vertices. The latter come weighted with two independent coupling constants. Once we pass to string theory, only one fundamental coupling remains. In fact, as illustrated in the right-hand side of Figure 1.3, any string theory diagram, no matter how many external legs and loops it has, may be cut into 3-vertices. Consequently, all interactions between strings are controlled by a single coupling constant g_s that comes with the 3-vertex depicted in Figure 1.2.

String theory possesses a consistent set of rules and elaborate computational tools to calculate scattering amplitudes. These produce formulas of the form (1.7).

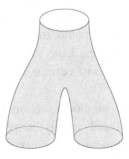

Figure 1.2 The only fundamental vertex of string theory describes the splitting and joining of strings.

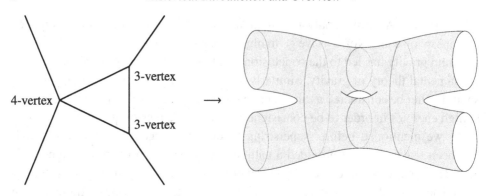

Figure 1.3 While point particles can posses many different types of vertices, all scattering amplitudes in string theory are composed of one fundamental 3-vertex.

It is of particular interest to study their low-energy properties. When $E \ll l_s^{-1}$, vibrational modes cannot be excited, and all we see are massless point-like objects. One may ask whether these behave like any of the particles we know. The answer is well known: at low energies, massless closed string modes scatter like gravitons [58, 77, 78] and a bunch of other particles that form the particle content of 10-dimensional supergravity theories. This observation is fundamental for string theory's advance into quantum gravity. In fact, it presents string theory as a consistent high-energy completion of gravitational theories.

1.3 Solitons and D-branes

For a moment, let us turn our attention to (super-)gravity theories. We are all familiar with the SCHWARZSCHILD solution of EINSTEIN's theory of gravity. It describes a black hole in our 4-dimensional world, i.e., a heavy object that is localized somewhere in space. Similar solutions certainly exist for the supergravity equations of motion. The massive (and charged) objects they describe may but need not be point-like localized in the 9-dimensional space. In fact, explicit solutions are known in which the mass density is localized along p-dimensional surfaces with $p = 1$ corresponding to strings, $p = 2$ to membranes, etc.; see [21] for a review. Such solutions were named black p-branes. Like ordinary black holes, however, most of these objects decay. But there exist certain extremal solutions, also known as solitonic p-branes, that are stable.

Now let us recall from the previous section that supergravity emerges as a low-energy description of closed string theory. Consequently, if supergravity contains massive $(p + 1)$-dimensional objects, the same should be true for closed string theory. One may therefore begin to wonder about the role p-branes could play in string theory. In order to gain some insight, let us suppose that a brane has been placed into the 9-dimensional space of our string background.

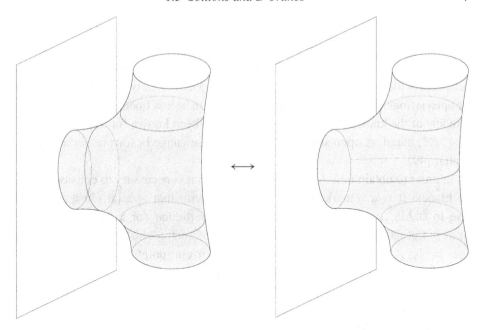

Figure 1.4 There are two ways to think about the interaction between massive closed string modes and D-branes, either in terms of closed string exchange or in terms of open string excitations.

Since it is heavy and charged, it will interact with the closed string modes in this background. In supergravity, we would describe this interaction through the exchange of gravitons or other particles mediating the relevant interaction. In string theory, a similar picture is possible only that now the interaction is mediated by exchange of closed strings as shown on the left-hand side of Figure 1.4.

But the figure suggests another way to think about the very same process. In order to allow for an unbiased view, we have re-drawn the interaction process on the right-hand side of Figure 1.4. What we see now is an infalling closed string that seems to open up when it hits the brane. For a brief period, an excited state is formed in which an open string propagates with both its ends remaining attached to the brane. Finally, this state decays again by emitting a closed string. Hence, we found two very different ways to think about exactly the same process. One of them involves an excited state of the p-brane in which an open string travels along the $p + 1$-dimensional world-volume. In order for such a state to exist, branes in string theory must be objects on which open strings can end. This is indeed the defining feature of so-called DIRICHLET p-branes or for short Dp-branes in string theory (see, e.g., [54]).

1.4 D-branes and Gauge Theory

In the previous section we argued that D-brane excitations can be thought of as open strings whose endpoints move within the p-dimensional space of a brane. Therefore, branes provide us with a second set of light objects, namely the vibrational modes of open strings. One can ask again whether the massless open string modes behave like any of the known particles. The answer has been known for a long time: When $E \ll l_s^{-1}$, massless open string modes scatter like gauge bosons or certain types of matter [49].

In order to obtain non-Abelian gauge theories it is necessary to consider clusters of branes. It is a remarkable fact of supergravity that special clusters can give rise to stable configurations. This is true in particular for a stack of N parallel branes. Let us enumerate the member branes of such a cluster or stack by indices $a, b = 1, \ldots, N$. Open strings must have their end-points moving along one of these N branes (see Figure 1.5). Since an open string has two ends, modes of an open string carry a pair a, b of "color" indices. Hence, massless open string modes on a stack of N parallel branes can be arranged in a $N \times N$ matrix, just as the components of a $U(N)$ YANG-MILLS field. In addition to non-Abelian gauge bosons, various matter multiplets can emerge from open strings. The precise matter content of the resulting low-energy theories depends much on the brane configuration under consideration, and we shall not make the attempt to describe it in any more detail.

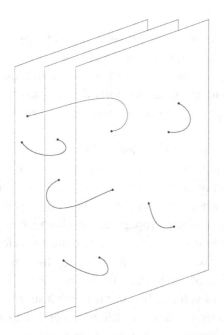

Figure 1.5 A stack of N branes with open strings stretched between them.

It is worth rehashing how $(p + 1)$-dimensional gauge theories have entered the stage through the back door. When we began this short cartoon of string theory, closed strings (and therefore gravitons) were all we had. Then we convinced ourselves that the theory contains additional heavy $(p+1)$-dimensional D-branes. Their excitations brought open strings into the picture and thereby another set of light degrees of freedom, including non-Abelian gauge bosons. Let us stress once more that the latter do not propagate in the 10-dimensional space-time but rather on the $(p+1)$-dimensional brane worlds. The dimension $p+1$ can take various values, one of them being $p + 1 = 4$. Our sketch of modern string theory has now brought us to the mid-1990s. At this point we have gathered all the ingredients that are necessary to discover a novel set of equivalences between gauge and string theory.

1.5 Gauge/String Dualities

The main origin of the novel dualities is not too difficult to grasp if we cleverly combine what we have seen in the previous section. To this end, let us suppose that we have placed two branes in our 10-dimensional background and that they are separated by some distance Δy. Since all branes are massive and charged objects, they will interact with each other. In supergravity, we would understand this interaction as an exchange of particles, such as gravitons, etc. Our branes, however, are objects in string theory, and hence there exists an infinite tower of vibrational closed string modes that mediate the interaction between them. A tree-level exchange is shown on the left-hand side of Figure 1.6. But as in our previous discussion, there exists another way to think about exactly the same process in terms of open strings. This is visualized on the right-hand side of Figure 1.6. There, the interaction appears to originate from pair creation/annihilation of open string modes with one end on each

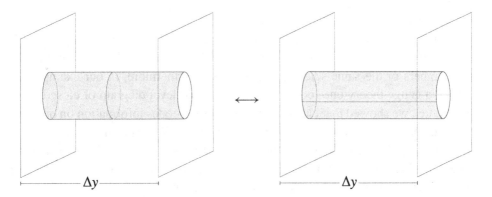

Figure 1.6 There are two ways to think about the interaction between two branes. One involves open strings, the other is mediated by closed string modes.

of the branes. In string theory, these two descriptions of the interaction give exactly the same final result for the force between the two branes.

A closer look reveals that the equivalence of our two computational schemes, one in terms of closed strings and the other in terms of open strings, is surprisingly non-trivial. Suppose, for example, that the distance Δy between the branes is very large. Then the closed string modes have to propagate very far in order to get from one brane to the other. Consequently, contributions from massive string modes may be neglected, and it is sufficient to focus on massless closed string modes, i.e., on the particles found in 10-dimensional supergravity. In the other regime in which the separation between the two branes becomes of the order of the string length l_s, such an approximation cannot give the right answer. Instead, the full tower of closed string modes must be taken into account. In other words, when $\Delta y \sim l_s$ the supergravity approximation breaks down, and we have to carry out a full string theory computation. From the point of view of open strings, the situation is reversed. When the branes are far apart, pair-created open strings propagate only briefly before they annihilate again, and hence the entire infinite tower of open string modes contributes to this computation. In the opposite regime where $\Delta y \sim l_s$, however, the interaction may be approximated by restricting ourselves to massless open string modes; i.e., all we need to perform is some gauge theory computation.

As simple as these comments on Figure 1.6 may seem, they suggest a remarkable conclusion: in the regime $\Delta y \sim l_s$, some calculation performed in the gauge theory on the world volume of our branes should lead to the same result as a full-fledged string theory calculation for closed strings propagating in the 10-dimensional background. A few aspects of this relation deserve to be stressed. In fact, we observe that it

- does preserve neither character nor difficulty of the computation,
- relates diagrams involving a different number of loops, and
- relates two theories in different dimensions, i.e., it is holographic.

These three features emerge clearly from our analysis. The first point is obvious. In fact, the two computations are so different that they would usually not be performed by members of the same scientific community. Furthermore, in our example, we related a gauge theory one-loop amplitude to a tree-level diagram of closed string theory, i.e., we showed that classical string theory encodes information on quantum gauge theory and vice versa. Finally, the gauge theory degrees of freedom are bound to the $(p+1)$-dimensional world-volume of our branes, whereas closed strings can propagate freely in 9+1 dimensions. We shall see these three features re-emerge in the concrete incarnations of the gauge/string theory dualities that we will discuss in Part II of this book.

Part I

Strings in Flat Backgrounds

The first part of this book deals with strings that move in flat MINKOWSKI space. While these string theories do not describe nature, their study shall allow us to illustrate many key concepts and constructions. In particular, we will be able to explore the precise relation between string theory and field theory on the world-sheet, the quantization of both closed and open strings, and the construction of consistent superstring theories. Along the way we will learn how to perform basic computations of scattering amplitudes and become quite fluent in switching between world-sheet and space-time interpretation of these calculations.

2

The Classical Closed Bosonic String

In this chapter we shall study in detail the classical description of closed bosonic strings. We will see and investigate three equivalent formulations. The third one shall later become our point of departure for the quantization of bosonic string theory. The presentation is essentially self-contained, but the reader may wish to consult some standard textbooks on classical mechanics for more background, such as [43] or the first few chapters of [35].

2.1 The Relativistic Particle

To begin, it is useful to revisit the classical description of free relativistic particles in the D-dimensional MINKOWSKI space. As usual, the motion of a relativistic point particle is governed by an action principle. It states that a particle moves along world-lines of minimal length. In order to compute the latter, we choose some parametrization $X(\tau) = (X^\mu(\tau), \mu = 0, \ldots, D-1)$, as shown in Figure 2.1, and write

$$S_1[X] = -m \int ds = -m \int d\tau \sqrt{-\dot{X}^\mu \dot{X}^\nu \eta_{\mu\nu}}. \qquad (2.1)$$

Here, η denotes the standard MINKOWSKI metric in a flat D-dimensional space $\mathcal{M}^D = \mathbb{R}^{1,D-1}$. The variation of this action w.r.t. the dynamical variables $X^\mu(t)$ yields the equations of motion for the free relativistic point particle

$$\frac{d}{d\tau} \frac{\dot{X}^\mu}{\sqrt{-\dot{X}^2}} = 0, \qquad (2.2)$$

where we introduced the notation $\dot{X}^2 = \dot{X}^\mu \dot{X}^\nu \eta_{\mu\nu}$. The action (2.1) nicely encodes the geometric principle behind the motion of point particles, but it has two important drawbacks. First of all, it can be used only for massive particles. In fact, if we

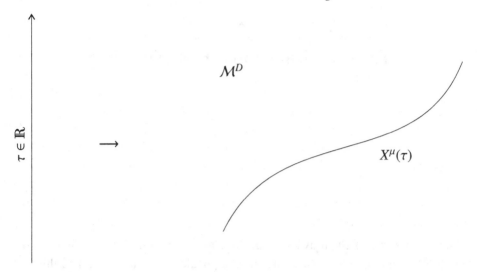

Figure 2.1 The world-line of a point particle can be parametrized by a map $X(\tau) = (X^{\mu}(\tau), \mu = 0,\ldots,D-1)$ from the real line into the D-dimensional space-time manifold.

set $m = 0$, then the action becomes trivial. The other concern one might have is that $S_1[X]$ leads to surprisingly difficult, non-linear equations of motion.

We can actually make progress on both issues by passing to a second description of point particles that – as we shall show momentarily – is equivalent to the first when $m \neq 0$. The new action S_2 for point particle motion involves one additional dynamical real variable $\gamma = \gamma(\tau)$, and it is constructed by

$$S_2[\gamma, X] = -\frac{1}{2} \int d\tau \, \sqrt{-\gamma} \, (\gamma^{-1}\dot{X}^2 + m^2) \,. \tag{2.3}$$

The way γ appears in this functional shows that we can consider it as an internal metric on the world-line. Variation w.r.t. γ and X^{μ} results in the following equations of motion:

$$\gamma^{-1}\dot{X}^2 - m^2 = 0 \,, \tag{2.4a}$$

$$\frac{d}{d\tau} \frac{\dot{X}^{\mu}}{\sqrt{-\gamma}} = 0 \,. \tag{2.4b}$$

In the massive case, the solution of eq. (2.4a) can be easily derived to be

$$\gamma(\tau) = \frac{\dot{X}^2}{m^2} \,. \tag{2.5}$$

Upon insertion of this formula into the second equation of motion (2.4b) we recover the non-linear equation (2.2) that was obtained from the first action principle. This shows that S_1 and S_2 are equivalent as long as $m \neq 0$. As we remarked before, S_1 does not provide a sensible theory if we set $m = 0$, since the action would be zero altogether. In contrast, the action S_2 remains non-trivial for $m = 0$. From now on, we shall focus on the motion of massless particles, i.e., on S_2 with $m = 0$.

In order to proceed to the third formulation, we observe that the action (2.3) has an interesting symmetry, namely that it is invariant under arbitrary reparametrizations $\varphi : \mathbb{R} \to \mathbb{R}$ of the world-line. More precisely, let us define the action of φ on our dynamical variables γ and X^μ through

$$\gamma(\tau) \mapsto \gamma_\varphi(\tau) = \gamma(\varphi(\tau))\dot{\varphi}(\tau)^2 , \tag{2.6}$$

$$X^\mu(\tau) \mapsto X^\mu_\varphi(\tau) = X^\mu(\varphi(\tau)) . \tag{2.7}$$

It is then easy to show that the action functional (2.3) enjoys the following invariance property:

$$S_2[\gamma_\varphi, X_\varphi] = S_2[\gamma, X] . \tag{2.8}$$

Hence, minima of S_2 come in infinite dimensional families that are related by simple reparametrizations. It is not hard to see that within each such family of solutions we can always find one representative that satisfies $\gamma = -1$. If we insert this particular choice of γ into our action (2.3), it reduces to

$$S_3[X] = \frac{1}{2} \int \mathrm{d}\tau \dot{X}^2. \tag{2.9}$$

Variation w.r.t. X^μ results in the following very simple linear equations of motion:

$$\ddot{X}^\mu = 0. \tag{2.10}$$

But this cannot quite be the whole story. Note that any straight line in the D-dimensional MINKOWSKI space solves eq. (2.10), including space-like ones. On the other hand, we know that massless particles move along null-like trajectories. What has happened is that in passing from S_2 to S_3 we have entirely ignored the equation of motion (2.4a). But the latter represents an important constraint on the allowed classical solutions

$$\dot{X}^2 = 0. \tag{2.11}$$

In formulating the equivalence between S_2 and S_3, this constraint must be taken into account. Solutions of S_2 are related to those solutions of S_3 that satisfy the constraint equation (2.11). Other solutions of S_3 exist, but they are not physically relevant.

It is useful to convince ourselves that the last description of particle motion as a constrained dynamical system involves the correct number of parameters.

The phase space for a relativistic point particle in a D-dimensional space-time is $2(D - 1)$-dimensional. As coordinates one may choose the $D - 1$ position coordinates of a particle at $\tau = 0$ along with the $(D - 1)$-dimensional initial momentum vector. On the other hand, straight lines in MINKOWSKI space possess $2D$ parameters. One of them is eliminated by the constraint (2.11). Furthermore, by shifts of the variable τ we can always achieve, e.g., that $X^0(0) = 0$. This leaves us with the correct number $2(D - 1)$ of phase space dimensions. In a more formal description, we start from a $2D$-dimensional phase space that is parametrized by the D coordinates x^μ and the same number of conjugate momenta p^μ. These satisfy the usual canonical POISSON commutation relation

$$\{p^\mu, x^\nu\} = \eta^{\mu\nu} \ .$$

The function $l = p^2 = \dot{X}^2$ gives rise to a first class constraint. As is well known, any such constraint generates a symmetry of the classical theory. In the case at hand,

$$\{l, X^\mu\} = 2p^\mu = 2\partial_\tau X^\mu \ ,$$

i.e., the constraint generates translations of τ. Thereby we can use it to eliminate two phase space coordinates, one through the symmetry it generates and the other by imposing the constraint equation $l = 0$.

2.2 The Relativistic String

Our discussion of the classical closed bosonic string will now proceed exactly along the same lines as in the particle models, i.e., we will start from some simple geometric principle and then reformulate the action in two steps until we end up with a constrained theory whose equations of motion are linear in the dynamical variables. In complete analogy to the point particle case we shall require that closed bosonic strings move along world-surfaces in the D-dimensional MINKOWSKI space whose area is minimal. We can compute the area of some surface by choosing a parametrization, i.e., a set of functions $(X^\mu(\tau, \sigma); \mu = 0, \ldots, D - 1)$ from a 2-dimensional cylinder $\mathbb{R} \times S^1$ into space-time. The coordinates on our cylinder are denoted by $\tau = \sigma^0$ and $\sigma = \sigma^1$. According to the usual recipes (see, e.g., chapter 6 of [80] for a very basic explanation), we may calculate the area of the world-surface by

$$S_{\mathrm{NG}}[X] = -\frac{1}{2\pi\alpha'} \int d\sigma \, d\tau \, \sqrt{-\det h_{ab}} \ . \tag{2.12}$$

This action is known as NAMBU-GOTO-action. It involves GRAM's determinant $h = \det h_{ab}$ of the induced metric h:

$$h_{ab} = \partial_a X^\mu \partial_b X^\nu \eta_{\mu\nu} \ . \tag{2.13}$$

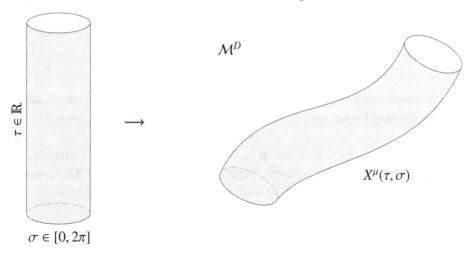

$\mathcal{M}^D$

$\tau \in \mathbb{R}$

$X^\mu(\tau, \sigma)$

$\sigma \in [0, 2\pi]$

Figure 2.2 The world-surface of a closed string is parametrized by map from the cylinder $\Sigma = \mathbb{R} \times [0, 2\pi]$ into the D-dimensional space-time manifold.

In all these formulas a, b assume the values $a, b = 0, 1$ while μ, ν run from $\mu, \nu = 0$ to $\mu, \nu = D-1$. The equations of motion for the NAMBU-GOTO-action are certainly highly non-linear. Therefore, we try once more to reformulate our dynamical principle, hoping that we can eventually relate it to some linear equations.

To this end, we now introduce a new action that involves an additional dynamical metric $\gamma_{ab}(\sigma, \tau)$ on the world-sheet $\mathbb{R} \times S^1$. Note that the real symmetric 2×2 matrix (γ_{ab}) contains three independent functions. The so-called POLYAKOV action involves the fields X^μ and γ_{ab},

$$S_{\text{Pol}}[\gamma, X] = -\frac{1}{4\pi\alpha'} \int d\sigma \, d\tau \, \sqrt{-\gamma} \gamma^{ab} \partial_a X^\mu \partial_b X^\nu \eta_{\mu\nu}, \qquad (2.14)$$

where we use $\gamma^{ab}(\sigma, \tau)$ to denote the matrix elements of the inverse metric, as usual, and we also introduced the shorthand $\gamma = \det(\gamma_{ab})$. The idea to include the world-sheet metric γ into the string theory action is actually due to [12, 17]. It was later exploited by Polyakov to perform the path integral quantization of strings; see, e.g., [55] for a detailed description. Variation with respect to the metric components using the simple relation $\delta\gamma = -\gamma\gamma_{ab}\delta\gamma^{ab}$[1] gives the equation

$$h_{ab} = \frac{1}{2}\gamma_{ab}\gamma^{cd}h_{cd}. \qquad (2.15)$$

---

[1] The variation of γ may be computed directly from LAPLACE's formula for the determinant.

We can divide this equation by the square root of its determinant to determine $\sqrt{-\gamma}\gamma^{ab}$ in terms of h^{ab}:

$$(-\gamma)^{\frac{1}{2}}\gamma^{ab} = h^{ab}(-h)^{\frac{1}{2}}. \tag{2.16}$$

Upon insertion into the POLYAKOV action (2.14) we are able to eliminate its dependence on the world-sheet metric and end up with the NAMBU-GOTO action. Thereby we have checked the equivalence of both descriptions.

Once more, the POLYAKOV action (2.14) enjoys a number of symmetries. To begin, it is invariant under *reparametrizations* on the world-sheet. The latter are given by maps $\varphi = (\varphi_a) : \mathbb{R} \times S^1 \longrightarrow \mathbb{R} \times S^1$ and act according to

$$\gamma_{ab}(\tau,\sigma) \mapsto \gamma_{ab}^{\varphi}(\tau,\sigma) = \frac{d\varphi_a}{d\sigma_c}\gamma_{cd}(\varphi(\tau,\sigma))\frac{d\varphi_b}{d\sigma_d}, \tag{2.17a}$$

$$X^{\mu}(\tau,\sigma) \mapsto X_{\varphi}^{\mu}(\tau,\sigma) = X^{\mu}(\varphi(\tau,\sigma)). \tag{2.17b}$$

We recall our convention that $\sigma^0 = \tau$ and $\sigma^1 = \sigma$. Such reparametrizations are determined by only two functions φ^a, and hence they are not sufficient to trivialize all three independent components of the world-sheet metric γ_{ab}. Fortunately, the POLYAKOV action possesses an additional symmetry under so-called WEYL *rescalings*. The latter involve a real function $\omega(\tau,\sigma)$ that leaves the parametrization X^{μ} of the world-surface invariant and acts on the world-sheet metric γ according to

$$\gamma_{ab}(\tau,\sigma) \mapsto \gamma_{ab}^{\omega}(\tau,\sigma) = \exp(2\omega(\tau,\sigma))\gamma_{ab}(\tau,\sigma). \tag{2.18}$$

It is not difficult to see that these symmetries of the POLYAKOV action under reparametrizations and WEYL rescalings suffice to bring the world-sheet metric into the standard form

$$\gamma_{ab} = \eta_{ab} = \begin{pmatrix} -1 & 0 \\ 0 & 1 \end{pmatrix}. \tag{2.19}$$

In this particular gauge, the POLYAKOV action assumes the form of an action for D free bosonic fields on the cylinder:

$$S_{\text{FF}}[X] = -\frac{1}{4\pi\alpha'}\int d\tau\, d\sigma\, \eta^{ab}\partial_a X^{\mu}\partial_b X^{\nu}\eta_{\mu\nu}. \tag{2.20}$$

As in the case of point particles, this action principle must be accompanied by the constraint equations

$$\partial_a X_{\mu}\partial_b X^{\mu} = \frac{1}{2}\eta_{ab}\partial_c X_{\mu}\partial^c X^{\mu}, \tag{2.21}$$

which follow from the equations (2.15) upon insertion of our special gauge choice $\gamma_{ab} = \eta_{ab}$. This means that only those solutions of the free field theory are admitted as physical solutions of the classical bosonic string that satisfy eq. (2.21).

2.3 Classical Bosonic Free Fields

The formulation of the classical bosonic string in terms of eqs. (2.20) and (2.21) will provide a basis for the quantization in the next few chapters. It is therefore useful to describe the phase space of the system in some more detail. From the classical action (2.20) we can read off the following equations of motion:

$$\partial_\tau^2 X^\mu - \partial_\sigma^2 X^\mu = 0 \,. \tag{2.22}$$

This means that all D components of $X = (X^\mu)$ solve the 2-dimensional wave equation on the cylinder. Let us recall that any solution of this simple partial differential equation can be written as a superposition of a left- and a right-moving term, i.e., it can be presented in the form $X^\mu(\tau,\sigma) = X^\mu(\sigma^+) + X^\mu(\sigma^-)$ where $\sigma^\pm$ denote the so-called light-cone coordinates $\sigma^\pm = \tau \pm \sigma$. Since we shall frequently switch between light-cone coordinates and our space $\sigma = \sigma^1$ and time $\tau = \sigma^0$ variables, let us use the occasion to list some of the relevant formulas:

$$\tau = \sigma^0 = \frac{1}{2}(\sigma^+ + \sigma^-), \qquad \sigma = \sigma^1 = \frac{1}{2}(\sigma^+ - \sigma^-), \tag{2.23}$$

$$\partial_\tau = \partial_+ + \partial_-, \qquad \partial_\sigma = \partial_+ - \partial_-, \tag{2.24}$$

$$d\tau \wedge d\sigma = d\sigma^0 \wedge d\sigma^1 = \frac{1}{2}d\sigma^- \wedge d\sigma^+ = 2d\sigma_- \wedge d\sigma_+, \tag{2.25}$$

where we introduced $\partial_\pm = \partial_{\sigma^\pm}$. The world-sheet metric η in light-cone coordinates is easily seen to take the form $\eta^{+-} = \eta^{-+} = -2$ and $\eta^{++} = 0 = \eta^{--}$. Hence, raising and lowering the light-cone indices $\pm$ brings in additional factors,

$$v^- = -2v_+ \quad , \quad v^+ = -2v_-,$$

for any vector v. This concludes our discussion of the coordinate change.

In the new light-cone coordinates the most general solution of eq. (2.22) on the cylinder is given by

$$X^\mu(\sigma^+, \sigma^-) = x^\mu + \alpha' p^\mu \tau + i\sqrt{\frac{\alpha'}{2}} \sum_{n \neq 0} \frac{1}{n}\left(a_n^\mu e^{-in\sigma^-} + \bar{a}_n^\mu e^{-in\sigma^+}\right), \tag{2.26}$$

where x^μ, p^μ, a_n^μ, and $\bar{a}_n^\mu$ are parameters of the solution that are fixed e.g. by an initial condition at $\tau = 0$. In order for X^μ to be real, we demand that

$$x^\mu = (x^\mu)^*, \quad p^\mu = (p^\mu)^*, \quad a_n^\mu = \left(a_{-n}^\mu\right)^*, \quad \bar{a}_n^\mu = \left(\bar{a}_{-n}^\mu\right)^*. \tag{2.27}$$

The variables x^μ and p^μ may be interpreted as the center of mass position and momentum of the string. Excitations are parametrized by two sets of independent oscillators that create left- and right-moving vibrations of the string. For convenience one often uses the dimensionless modes a_0^μ and $\bar{a}_0^\mu$ that are related to the momentum p^μ by

$$a_0^\mu := \sqrt{\frac{\alpha'}{2}} p^\mu =: \bar{a}_0^\mu . \tag{2.28}$$

The phase space of our field theory may be parametrized by the fields $X^\mu(\tau, \sigma)$ and the corresponding conjugate momenta

$$\Pi^\mu(\tau, \sigma) = \frac{1}{2\pi\alpha'} \dot{X}^\mu(\tau, \sigma) \tag{2.29}$$

evaluated at some fixed time $\tau = \tau^0$. This space comes equipped with the following canonical POISSON structure:

$$\left\{ X^\mu(\tau, \sigma); X^\nu(\tau, \sigma') \right\}_{\text{P.B.}} = \left\{ \dot{X}^\mu(\tau, \sigma); \dot{X}^\nu(\tau, \sigma') \right\}_{\text{P.B.}} = 0 , \tag{2.30a}$$

$$\left\{ \dot{X}^\mu(\tau, \sigma); X^\nu(\tau, \sigma') \right\}_{\text{P.B.}} = 2\pi\alpha' \eta^{\mu\nu} \delta(\sigma - \sigma'). \tag{2.30b}$$

Alternatively, we can think of the phase space as being parametrized by x^μ, p^μ, and the FOURIER modes $a_n^\mu, \bar{a}_n^\mu$. For these coordinate functions on the phase space, the POISSON structure reads

$$\left\{ a_n^\mu; a_m^\nu \right\}_{\text{P.B.}} = i\, n\, \eta^{\mu\nu} \delta_{n+m,0} = \left\{ \bar{a}_n^\mu; \bar{a}_m^\nu \right\}_{\text{P.B.}} , \tag{2.31a}$$

$$\left\{ a_n^\mu; \bar{a}_m^\nu \right\}_{\text{P.B.}} = 0 , \tag{2.31b}$$

$$\left\{ x^\mu; a_n^\nu \right\}_{\text{P.B.}} = 0 = \left\{ x^\mu; \bar{a}_n^\nu \right\}_{\text{P.B.}} \quad for \quad n \neq 0 , \tag{2.31c}$$

$$\left\{ p^\mu; x^\nu \right\}_{\text{P.B.}} = \eta^{\mu\nu} . \tag{2.31d}$$

It is rather instructive to re-derive the equal time POISSON brackets (2.30) for the fields X and $\dot{X}$ from their mode expansion (2.26) and the POISSON brackets (2.31).

2.4 Stress Energy Tensor and Constraints

In this final section we would like to study properties of the stress energy tensor of our free bosonic field theory. According to the usual rules, we can compute its components from the LAGRANGE density $\mathcal{L}$ by[2]

$$T_{ab} = -2\pi \frac{\delta\mathcal{L}}{\delta\partial^a X^\mu} \partial_b X^\mu + 2\pi \eta_{ab}\mathcal{L} = -\frac{1}{\alpha'} \partial_a X^\mu \partial_b X_\mu + \frac{1}{2\alpha'} \eta_{ab} \partial_c X^\mu \partial^c X_\mu. \tag{2.32}$$

---

[2] Following standard conventions in string theory we have included an additional factor of -2π in the definition of the stress energy tensor T.

The stress energy tensor of our free field theory enjoys a number of important fundamental properties that may easily be checked in the case at hand:

1. The stress energy tensor T_{ab} is symmetric, i.e., $T_{ab} = T_{ba}$.
2. The stress energy tensor T_{ab} is conserved, i.e., $\partial_a T^a{}_b = 0$.
3. The stress energy tensor T_{ab} is traceless, i.e., $T^a{}_a = 0$.

While the first two properties hold in any relativistic field theory, the last one is characteristic for models that do not contain any fundamental scale, i.e., for massless field theories.

As they stand, the properties of the tensor T_{ab} are valid in any coordinate system on the 2-dimensional world-sheet. It is particularly instructive to spell them out in the light-cone coordinates $\sigma^\pm = \tau \pm \sigma$ that we have already introduced above. In these coordinates

1'. The symmetry property 1 of the stress energy tensor reads $T_{+-} = T_{-+}$.
3'. Tracelessness becomes $T_{-+} + T_{+-} = 0$, which, together with the symmetry property 1', implies that $T_{+-} = T_{-+} = 0$.
2'. The conservation law 2 of T reads $\partial_- T_{++} = \partial_+ T_{--} = 0$.

In conclusion, we have shown that the stress energy tensor has only two non-vanishing components, namely, $T_{\pm\pm}$. Moreover, each of them depends merely on one of the light-cone variables. We can exploit the second fact to introduce the modes l_n and $\bar{l}_n$ of the stress energy tensor through

$$T_{--}(\sigma^-) = -\sum_{n \in \mathbb{Z}} l_n e^{-in\sigma^-} \quad \text{with} \quad l_n = \frac{1}{2} \sum_{m \in \mathbb{Z}} a_m^\mu a_{n-m}^\nu \eta_{\mu\nu}, \qquad (2.33a)$$

$$T_{++}(\sigma^+) = -\sum_{n \in \mathbb{Z}} \bar{l}_n e^{-in\sigma^+} \quad \text{with} \quad \bar{l}_n = \frac{1}{2} \sum_{m \in \mathbb{Z}} \bar{a}_m^\mu \bar{a}_{n-m}^\nu \eta_{\mu\nu}. \qquad (2.33b)$$

With the help of the fundamental POISSON commutators (2.31) for the modes of X^μ we can determine the POISSON algebra of the modes l_n and $\bar{l}_n$:

$$\{l_n; l_m\}_{\text{P.B.}} = i(n - m) l_{n+m}, \qquad (2.34a)$$

$$\{\bar{l}_n; \bar{l}_m\}_{\text{P.B.}} = i(n - m) \bar{l}_{n+m}. \qquad (2.34b)$$

Moreover, the modes l_n of T_{--} POISSON commute with the modes $\bar{l}_n$ of T_{++}. It is also interesting to note that

$$\{l_n; X^\mu(\sigma^+, \sigma^-)\}_{\text{P.B.}} = i e^{in\sigma^-} \partial_{\sigma^-} X^\mu(\sigma^+, \sigma^-). \qquad (2.35)$$

A similar relation with σ^+ instead of σ^- holds for the POISSON commutator with $\bar{l}_n$. This means that the modes l_n and $\bar{l}_n$ generate diffeomorphisms of the light-cone coordinates σ^- and σ^+, respectively.

To conclude this chapter we recall that passing from free field theory to bosonic string theory involves imposing an additional constraint. A brief comparison of the relevant formula (2.21) with the stress energy tensor (2.32) shows that the constraints of bosonic string theory may be rephrased as $T_{ab} = 0$. In terms of the modes l_n and $\bar{l}_n$, they take the simple form

$$l_n = \bar{l}_n = 0 \qquad for \ n \in \mathbb{Z}. \tag{2.36}$$

Put differently, we descend from the simple phase space of free bosonic field theory to the phase space of bosonic string theory by imposing the infinite number of constraint equations (2.36). We have shown before that the constraint functions l_n and $\bar{l}_n$ possess a closed POISSON algebra (2.34a). In more fancy terms, the constraints we are imposing are of first class. As we recalled a bit earlier, first class constraints generate gauge symmetries. In the case at hand, these are simply the reparametrizations of light-cone coordinates. Each constraint thereby eliminates two degrees of freedom, one through the symmetry it generates, the other through the constraint equation itself.

Exercises

Problem 1. *Derive the equations of motion (2.15) for the metric γ from the POLYAKOV action functional (2.14). In order to do so, also derive the expression for the variation $\delta\gamma$ that was stated in the text before eq. (2.15).*

Problem 2. *Derive the POISSON brackets (2.30) for the fields X^μ and $\dot{X}^\mu$ from the POISSON brackets (2.31) for the variables $a_n^\mu, \bar{a}_n^\mu$, and x^μ.*

Problem 3. *Use the POISSON brackets (2.31) for oscillators and the definition (2.33a) of the variables l_n to derive the WITT-POISSON algebra (2.34a).*

Problem 4. *Show that the elements l_n which were defined in eq. (2.33a) generate infinitesimal diffeomorphisms acting on the light-cone variable σ^+, i.e., that they satisfy eq. (2.35).*

Problem 5. *The action of a 2-dimensional free massive scalar field is given by*

$$S[X] = \frac{1}{g^2} \int d\tau d\sigma \left(\eta^{ab} \partial_a X \partial_b X - m^2 X^2 \right).$$

Construct the symmetric and conserved stress energy tensor $T = (T_{ab})$ and compute its trace.

3

Free Bosonic Quantum Field Theory

In this chapter we shall forget about string theory and simply quantize a free 2-dimensional bosonic field theory. Our aim is to construct its state space and the fields that act on it and then to compute correlation functions. Finally, we shall also study the properties of the stress energy tensor of this very simple quantum field theory.

3.1 Construction of the State Space

In the canonical quantization procedure the commutation relations of operators are obtained from the POISSON brackets by the prescription

$$\{\cdot\,;\,\cdot\}_{\text{P.B.}} \;\mapsto\; i[\,\cdot\,,\,\cdot\,]. \tag{3.1}$$

Through this replacement we obtain from the POISSON structure (2.31) the following commutation relations of the basic operators $a_n^\mu, \bar{a}_m^\mu, p^\mu$, and x^μ:

$$\left[a_n^\mu, a_m^\nu\right] = n\,\eta^{\mu\nu}\,\delta_{n+m,0} = \left[\bar{a}_n^\mu, \bar{a}_m^\nu\right], \tag{3.2a}$$

$$\left[a_n^\mu, \bar{a}_m^\nu\right] = 0, \tag{3.2b}$$

$$\left[x^\mu, a_n^\nu\right] = 0 = \left[x^\mu, \bar{a}_n^\nu\right] \quad \text{for} \quad n \neq 0, \tag{3.2c}$$

$$\left[x^\mu, p^\nu\right] = i\,\eta^{\mu\nu}. \tag{3.2d}$$

To be more precise, we should introduce new symbols for the operators of the quantum theory to distinguish them clearly from the coordinates of the classical phase space. But we trust that the context will always show whether we are talking about operators or their classical analogues.

Note that the pairs a_n^μ and a_{-n}^μ form an infinite set of oscillators with the usual non-vanishing commutator. We declare all operators a_n^μ and $\bar{a}_n^\mu$ with negative index $n < 0$ to be creation operators and then treat the remaining a_n^μ and $\bar{a}_n^\mu$ with positive index $n > 0$ as annihilation operators. By definition, the ground states $|k\rangle$ are annihilated by all annihilation operators, and they are eigenstates of the momentum operators p^μ:

$$a_n^\mu \,|k\rangle = 0 = \bar{a}_n^\mu \,|k\rangle \quad \text{for} \quad n > 0, \tag{3.3a}$$

$$p^\mu \,|k\rangle = k^\mu \,|k\rangle \,. \tag{3.3b}$$

From each ground state $|k\rangle$, we construct a component $\mathscr{H}_k$ of the full state space by acting with our creation operators:

$$\mathscr{H}_k = \left\{ a_{n_1}^{\mu_1} \cdots a_{n_r}^{\mu_r} \bar{a}_{m_1}^{\nu_1} \cdots \bar{a}_{m_s}^{\nu_s} \,|k\rangle \,\bigm|\, s, r \in \mathbb{N}_0; \, n_1 \leq \ldots \leq n_r \leq 0; \right.$$
$$\left. m_1 \leq \ldots \leq m_s \leq 0 \right\}. \tag{3.4}$$

The total state space of our free field theory is given by some kind of sum over all the values for k. Since k is continuous, the appropriate construction is that of a direct integral [20]:

$$\mathscr{H}_D^{\mathrm{CBF}} = \oint d^D k \, \mathscr{H}_k = \oint d^D k \, \mathcal{H}_k \otimes \bar{\mathcal{H}}_k \,. \tag{3.5}$$

In getting to the expression on the right we made use of the fact that each block $\mathscr{H}_k$ can be written as a product $\mathcal{H}_k \otimes \bar{\mathcal{H}}_k$ where, e.g., $\mathcal{H}_k$ is generated by unbared operators $a_n^\mu, n < 0$ only. Our superscript "CBF" stands for *closed bosonic field*.

The space $\mathscr{H}_D^{\mathrm{CBF}}$ comes equipped with a natural bilinear form that is uniquely specified by the normalization condition for ground states

$$\langle k \,|\, k' \rangle = \delta^{(D)}(k - k'), \tag{3.6}$$

along with the requirement that the associated operation $*$ of taking adjoints takes the form

$$\left(a_n^\mu \right)^* = a_{-n}^\mu \,, \quad \left(\bar{a}_n^\mu \right)^* = \bar{a}_{-n}^\mu \,. \tag{3.7}$$

Obviously, this is the quantum theoretical analogue of the classical reality conditions (2.27). In order to see how these requirements determine our bilinear form, let us evaluate

$$\left(a_{-1}^\mu |k'\rangle , \, a_{-1}^\mu |k\rangle \right) = \langle k' | \left(a_{-1}^\mu \right)^* a_{-1}^\mu | k \rangle = \langle k' | a_1^\mu a_{-1}^\mu | k \rangle$$
$$= \langle k' | \left[a_1^\mu, a_{-1}^\mu \right] | k \rangle = \eta^{\mu\mu} \langle k' | k \rangle$$
$$= \begin{cases} -\delta(k - k') & \text{for } \mu = 0, \\ +\delta(k - k') & \text{for } \mu \in \{1, \ldots, D - 1\}. \end{cases} \tag{3.8}$$

It is clear how similar computations can be performed for states with a larger number of creation operators. They certainly are slightly more difficult to perform.

There is one outcome of our short computation that deserves to be stressed: we should note that our bi-linear form is not positive definite. Put differently, it fails to turn the state space $\mathcal{H}_D^{\text{CBF}}$ into a HILBERT space. The origin of the problem is fairly obvious: it is directly linked to the existence of the time-like bosonic field X^0 for which the metric has a negative sign. We will have a lot more to say about this when we discuss the quantization of closed bosonic strings. For the moment, however, it is a simple fact to move along with.

3.2 Fields and State Field Correspondence

In the following it will be convenient to WICK rotate the time coordinate, i.e., to replace the coordinate τ by $-i\tau$. When this substitution is applied to the exponentials of our light-cone coordinates $\sigma^\pm = \tau \pm \sigma$, we find

$$
\begin{aligned}
e^{i\sigma^-} &= e^{i(\tau-\sigma)} \mapsto e^{\tau - i\sigma} =: z , \\
e^{i\sigma^+} &= e^{i(\tau+\sigma)} \mapsto e^{\tau + i\sigma} =: \bar{z} .
\end{aligned}
\tag{3.9}
$$

To be more precise, we should write $\sigma_e^\pm$ instead of $\sigma^\pm$ on the right side. But we will not use these $\sigma_e^\pm$ much and work with z and $\bar{z}$ instead, which are always assumed to be complex conjugate of each other. We want to list a few properties of the change of variables from $\sigma^\pm$ to z and $\bar{z}$. Denoting derivatives ∂_z with respect to z by $\partial_z = \partial$ and similarly introducing $\partial_{\bar{z}} = \bar{\partial}$ we have

$$
\partial_- = iz\partial, \qquad \partial_+ = i\bar{z}\bar{\partial} ,
\tag{3.10}
$$

$$
d\sigma^- = (iz)^{-1}dz, \qquad d\sigma^+ = (i\bar{z})^{-1}d\bar{z} .
\tag{3.11}
$$

More generally, for a quantity $A(\sigma^-, \sigma^+)$ of weight $h, \bar{h}$ one has

$$
A(\sigma^-, \sigma^+) = (iz)^h (i\bar{z})^{\bar{h}} A(z, \bar{z}) .
\tag{3.12}
$$

In the new coordinates, the quantum fields X^μ read

$$
X^\mu(z, \bar{z}) = x^\mu - \frac{i}{2}\alpha'p^\mu \ln(z\bar{z}) + i\sqrt{\frac{\alpha'}{2}} \sum_{n\neq 0} \left(\frac{a_n^\mu}{n} z^{-n} + \frac{\bar{a}_n^\mu}{n} \bar{z}^{-n} \right) .
\tag{3.13}
$$

Since the fields X^μ are objects of weight zero, the expression we displayed is simply obtained from eq. (2.26) by the change of variables $\sigma^- = \sigma^-(z) = -i\ln z$ and $\sigma^+ = \sigma^+(\bar{z}) = -i\ln\bar{z}$. In addition we have also replaced the classical variables by operators, i.e., we performed our canonical quantization. X^μ are the basic fields of our model, but there are a few others that will play an important role in the following. These include in particular the derivatives of X^μ, which take the form

$$J^\mu(z) := i \partial X^\mu(z, \bar{z}) = \sqrt{\frac{\alpha'}{2}} \sum_n a_n^\mu z^{-n-1} , \tag{3.14a}$$

$$\bar{J}^\mu(\bar{z}) := i \bar{\partial} X^\mu(z, \bar{z}) = \sqrt{\frac{\alpha'}{2}} \sum_n \bar{a}_n^\mu \bar{z}^{-n-1}. \tag{3.14b}$$

Let us stress that these fields are (anti-)holomorphic, i.e., they depend only on z or $\bar{z}$ but not on both. Fields with this property are known as *chiral*. While J^μ and $\bar{J}^\mu$ are straightforward to construct, a bit more thought is needed to define the so-called *vertex operators*:

$$V_k(z, \bar{z}) = \; : \exp\left(i k \cdot X(z, \bar{z})\right) :$$

$$= \exp\left(i k \cdot X_<(z, \bar{z})\right) \exp\left(i k \cdot X_>(z, \bar{z})\right) . \tag{3.15}$$

Exponentials of the bosonic fields X^μ are highly non-linear in the basic oscillators, and therefore their definition requires some normal ordering. We have denoted the latter by $: \cdot :$ and, as usual, agree that it instructs us to move all annihilators (the a_n^μ with $n \geq 0$) to the right of the creation operators (a_n^μ with $n < 0$ and x^μ). The outcome of this rule is made explicit in the second formula for V_k where we rewrote the vertex operators in terms of

$$X_<^\mu(z, \bar{z}) = x^\mu + i \sqrt{\frac{\alpha'}{2}} \sum_{n<0} \left(\frac{a_n^\mu}{n} z^{-n} + \frac{\bar{a}_n^\mu}{n} \bar{z}^{-n} \right) , \tag{3.16a}$$

$$X_>^\mu(z, \bar{z}) = -\frac{i}{2} \alpha' p^\mu \ln(z\bar{z}) + i \sqrt{\frac{\alpha'}{2}} \sum_{n>0} \left(\frac{a_n^\mu}{n} z^{-n} + \frac{\bar{a}_n^\mu}{n} \bar{z}^{-n} \right) . \tag{3.16b}$$

The fields V_k, J^μ, and $\bar{J}^\mu$ will serve as basic building blocks. Certainly we can construct further composite fields by multiplication and by taking derivatives.

The space of fields in our field theory has an interesting property that shall become very important later: it is isomorphic to the space of states. More precisely, for any state $| \psi \rangle \in \mathscr{H}^{\text{CBF}}$ there exists one and only one field $V_\psi(z, \bar{z})$ such that

$$V_\psi(z, \bar{z}) | 0 \rangle \big|_{(z,\bar{z})=(0,0)} = | \psi \rangle . \tag{3.17}$$

Before we specify the form of V_ψ for an arbitrary state $| \psi \rangle$ it is instructive to consider a few simple examples. Let us begin with the vertex operators V_k and evaluate

$$V_k(z, \bar{z}) \, | \, 0 \rangle \, |_{(z,\bar{z})=(0,0)} = \exp{(i \, k \cdot X_<(z, \bar{z}))} \exp{(i \, k \cdot X_>(z, \bar{z}))} \, | \, 0 \rangle |_{(z,\bar{z})=(0,0)}$$

$$= \exp{(i \, k \cdot X_<(z, \bar{z}))} \, | \, 0 \rangle |_{(z,\bar{z})=(0,0)} = \exp{(i \, k \cdot x)} \, | \, 0 \rangle = | \, k \rangle.$$

In the first step we have used that $\exp(ikX_>(z, \bar{z})) \, | \, 0 \rangle = | \, 0 \rangle$ for all z simply because every term in $X_>$ annihilates the vacuum $| \, 0 \rangle$. Once this is inserted, we can actually set $(z, \bar{z}) = (0, 0)$ because all terms involving negative powers of z and $\bar{z}$ have dropped out. We are then left with an action of $\exp(ikx)$ on the vacuum. From the commutation relations between p^μ and x^ν it is easy to infer that $\exp(ikx) \, | \, 0 \rangle = | \, k \rangle$, which proves our claim. Hence, we have shown that V_k is the field that corresponds to the ground state $| \, k \rangle$ in the sense of eq. (3.17). Next we wish to illustrate how excitations of the ground states can emerge. To this end we note that

$$\partial^n J^\mu(0) \, | \, 0 \rangle = n! \, \sqrt{\frac{\alpha'}{2}} \, a^\mu_{-1-n} \, | \, 0 \rangle. \tag{3.18}$$

The reasoning is again rather straightforward. Half of the terms in $\partial^n J^\mu$ annihilate the vacuum $| \, 0 \rangle$. The creation operators $a^\mu_m, m < -n$, which contribute to $\partial^n J^\mu(z)$, are multiplied with z^{-m-n-1}. Hence, there is only one term that does not vanish when we set $z = 0$: the one with $m = -n - 1$. Taking into account some additional numerical factors we arrive at eq. (3.18).

We are now well prepared to construct the state field correspondence in full generality. Suppose that we consider the state of the form

$$| \, \psi \rangle = \left(\frac{\alpha'}{2} \right)^{\frac{r+s}{2}} \prod_{c=1}^{r} (-n_c - 1)! \prod_{d=1}^{s} (-m_d - 1)! \, a^{\mu_1}_{n_1} \cdots a^{\mu_r}_{n_r} \bar{a}^{\nu_1}_{m_1} \cdots \bar{a}^{\nu_s}_{m_s} \, | \, k \rangle. \tag{3.19a}$$

Then the corresponding operator takes the form

$$V_\psi =: \partial^{-n_1-1} J^{\mu_1}(z) \cdots \partial^{-n_r-1} J^{\mu_r}(z) \bar{\partial}^{-m_1-1} \bar{J}^{\nu_1}(\bar{z}) \cdots \bar{\partial}^{-m_s-1} \bar{J}^{\nu_s}(\bar{z}) \exp{(i \, k \cdot X(z, \bar{z}))} :. \tag{3.19b}$$

In order to check our claim one has to show that V_ψ has the property (3.17). We leave this as an exercise.

3.3 Correlation Functions

Our next aim is to compute correlation functions involving an arbitrary number of fields V_ψ. We shall begin with correlators of the vertex operators V_k and then explain below how additional insertions of the chiral fields J^μ and $\bar{J}^\nu$ can be treated.

In order to compute the vacuum expectation value of a product of vertex operators our main task is to reorder the factors such that in the end all the annihilation operators stand to the right of the creation operators. The following formula shows how this works for a product of two vertex operators:

$$V_{k_1}(z, \bar{z}) V_{k_2}(w, \bar{w}) = |z - w|^{\alpha' k_1 \cdot k_2} \exp(i k_1 \cdot X_<(z, \bar{z}) + i k_2 \cdot X_<(w, \bar{w}))$$
$$\times \exp(i k_1 \cdot X_>(z, \bar{z}) + i k_2 \cdot X_>(w, \bar{w})) . \quad (3.20)$$

Once the definition of the vertex operators is plugged into the left-hand side of this equation, we have a product of four factors with an alternating appearance of annihilation and creation operators. Passing to the right-hand side requires us to move the annihilating factor of the first vertex operator past the creating factor of the second. This can be achieved with the help of the usual BAKER-CAMPBELL-HAUSDORFF formula. All we need to compute is the commutator between the creation and annihilation part of X:

$$[X_>(z, \bar{z}), X_<(w, \bar{w})] = -\frac{1}{2}\eta^{\mu\nu}\alpha' \ln(z\bar{z}) + \eta^{\mu\nu}\frac{\alpha'}{2} \sum_{n=1}^{\infty} \left(\frac{1}{n}\frac{w^n}{z^n} + \frac{1}{n}\frac{\bar{w}^n}{\bar{z}^n}\right) \quad (3.21)$$
$$= -\eta^{\mu\nu}\alpha' \ln|z - w|.$$

In the derivation we have inserted the definition (3.16a) and (3.16b) of $X_<$ and $X_>$, the commutation relations (3.2), and, finally, the expansion of the logarithm, i.e., $\ln(1 - x) = -\sum_1^{\infty} x^n/n$. If we were to compute the vacuum expectation value for a product of two vertex operators we would simply sandwich our formula (3.20) between the incoming and outgoing vacuum states. By the same arguments that we have employed before, properties of the vacuum remove the field-dependent factors so that we are left with

$$\langle V_{k_1}(z, \bar{z}) V_{k_2}(w, \bar{w}) \rangle = |z - w|^{\alpha' k_1 \cdot k_2} \langle k_1 + k_2 | 0 \rangle = |z - w|^{\alpha' k_1 \cdot k_2} \delta^{(D)}(k_1 + k_2).$$

An iteration of the same procedure leads to the following final result for the N-point correlator of vertex operators:

$$\langle V_{k_1}(z_1, \bar{z}_1) \cdots V_{k_n}(z_n, \bar{z}_n) \rangle = \prod_{i<j} |z_i - z_j|^{\alpha' k_i \cdot k_j} \delta^{(D)}\left(\sum_{i=1}^{n} k_i\right). \quad (3.22)$$

Correlation functions involving currents J^μ are determined using an analogous decomposition into a creation and an annihilation part:

$$J^\mu(z) = \underbrace{\sqrt{\frac{\alpha'}{2}} \sum_{n\geq 0} a_n^\mu z^{-n-1}}_{J_>^\mu(z)} + \underbrace{\sqrt{\frac{\alpha'}{2}} \sum_{n<0} a_n^\mu z^{-n-1}}_{J_<^\mu(z)} \quad (3.23)$$

along with the commutators

$$\left[J_{>}^{\mu}(z), J^{\nu}(w)\right] = \frac{\alpha'}{2} \frac{\eta^{\mu\nu}}{(z-w)^2},$$ (3.24a)

$$\left[J_{>}^{\mu}(z), V_k(w, \bar{w})\right] = \frac{\alpha'}{2} \frac{k^{\mu}}{z-w} V_k(w, \bar{w}).$$ (3.24b)

If the field J^{μ} appears somewhere in our correlation function, we split it according to eq. (3.23) and then move the annihilation part to the right until it hits the vacuum $|0\rangle$. Similarly, we move the creation part to the left. Along the way, we will have to commute $J_{>}^{\mu}$ and $J_{<}^{\mu}$ past all other currents J^{ν} and vertex operators V_k using the two commutation relations we have stated.

3.4 Stress Energy Tensor and the VIRASORO Algebra

Before we conclude this chapter, let us address once more the stress energy tensor of the theory and see how its properties are effected by the quantization. As we have explained before, T possesses only two independent components, T_{++} and T_{--}, which each depend on one light-cone coordinate only. In our new coordinates $z, \bar{z}$ the right-moving (holomorphic) component of the stress energy tensor is given by

$$T(z) := (iz)^{-2} T_{--}(\sigma^-(z)) = -\frac{1}{\alpha'} : \partial X^{\mu} \partial X_{\mu} := \frac{1}{\alpha'} : J^{\mu}(z) J_{\mu}(z) : .$$ (3.25)

As in the classical theory, we introduce the modes L_n of the stress energy tensor through a LAURENT expansion in the world-sheet coordinate z:

$$T(z) = \sum_{n \in \mathbb{Z}} L_n z^{-n-2}.$$ (3.26)

In terms of our basic oscillators, the quantities L_n read

$$L_n = \frac{1}{2} \sum_{n \in \mathbb{Z}} : a_{n-m}^{\mu} a_m^{\nu} : \eta_{\mu\nu}.$$ (3.27)

The central result of our discussion here concerns the commutator of the L_m. In fact, we claim that the modes $L_n = (L_{-n})^*$ of the stress energy tensor generate a VIRASORO algebra

$$[L_n, L_m] = (n-m)L_{n+m} + \frac{c}{12} n(n^2 - 1)\delta_{n+m,0}$$ (3.28)

with the so-called central charge $c = D$ given by the dimension of our MINKOWSKI background. Note that the second term on the right-hand side was not present in the classical limit (2.34a); see, e.g., [27] for a derivation of this term. The new term arises because of the normal ordering that is needed in the quantum theory in order to properly define the operators L_n. For $c = 0$, the relations (3.28) would define an infinite dimensional Lie algebra known as the WITT algebra. Adding the second term is referred to in mathematics as a *central extension* because the extending term commutes with all the L_n. As we have seen, the precise value of the central charge c depends on the background. It takes integer values for flat backgrounds, but can also assume non-integer values once we start to consider strings moving in curved background geometries.

For later use we would like to briefly list a few further properties of the VIRASORO generators L_n. To begin, let us note that L_n with positive index $n > 0$ annihilate the ground states

$$L_n \,|\, k \rangle = 0 \qquad for \quad n > 0, \tag{3.29a}$$

while the zero mode gives

$$L_0 \,|\, k \rangle = \frac{\alpha'}{4} k^2 \,|\, k \rangle \; =: \; h_k \,|\, k \rangle . \tag{3.29b}$$

Furthermore, the commutator with our basic oscillators can be evaluated to take the form

$$\left[L_n, a_m^\mu \right] = -m\, a_{n+m}^\mu . \tag{3.30}$$

From the last property of L_n one may infer the following commutation relation between L_n and our vertex operators V_k:

$$[L_n, V_k(z, \bar{z})] = z^n \left(z\partial_z + h_k(n+1) \right) V_k(z, \bar{z}) . \tag{3.31}$$

This relation has a nice interpretation. It shows that our VIRASORO elements L_n generate diffeomorphisms acting on the coordinate z and that under such diffeomorphisms the vertex operators transform as tensors of weight $h_k = \alpha' k^2/4$. Indeed, let us consider some diffeomorphism $\phi : z \mapsto \phi(z)$. Under the action of such a diffeomorphism, a tensor of weight h transforms according to

$$A(z) \mapsto A_\phi(z) = (\partial_z \phi(z))^h A(\phi(z)).$$

For an infinitesimal diffeomorphism of the form $\phi_n^\epsilon(z) = z + \epsilon z^{n+1}$ we obtain

$$\delta A := A_\phi(z) - A(z) = (1 + \epsilon(n+1)z^n)^h A(z + \epsilon z^{n+1}) - A(z)$$

$$= \epsilon \left(z^{n+1} \partial_z A(z) + h(n+1)z^n A(z) \right) + \ldots = \epsilon z^n \left(z\partial_z + h(n+1) \right) A .$$

Hence, the leading term in the variation of our tensor A agrees exactly with the right-hand side of eq. (3.31). The relation (3.31) is the quantum theory version of the classical POISSON commutator (2.35). All functions of X^μ possess a vanishing weight before quantization. The term involving h_k is a mere quantum effect that arises from normal ordering.

Exercises

Problem 6. *Show that the operators L_n that were defined in eq. (3.27) satisfy the commutation relations (3.28) of a* VIRASORO *algebra with central charge $c = D$ given by the dimension of* MINKOWSKI *space.*

Problem 7. *Show that the* VIRASORO *generators L_n that were defined in eq. (3.27) generate diffeomorphisms of the world-sheet variable z, i.e. that the vertex operators (3.15) satisfy commutation relations of the form (3.31).*

Problem 8. *For the currents $J^\mu(z)$ and the* VIRASORO *field $T(z)$ that were defined in eqs. (3.14a) and (3.25), respectively, compute the following correlation functions:*

1. $\langle 0|J^\mu(z)J^\nu(w)|0\rangle$,
2. $\langle 0|T(z)T(w)|0\rangle$,
3. $\langle 0|J^\mu(z_1)J^\nu(z_2)J^\rho(z_3)|0\rangle$.

Problem 9. *Compute the following correlation function*

$$\langle 0|J^\mu(w)V_{k_1}(z_1,\bar{z}_1)V_{k_2}(z_2,\bar{z}_2)V_{k_3}(z_3,\bar{z}_3)|0\rangle,$$

where the vertex operators $V_k(z,\bar{z})$ and the currents $J^\mu(w)$ are defined in eqs. (3.15) and (3.14a), respectively.

HINT: *Reduce the computation of the 4-point function to a 3-point function of vertex operators. The latter is given by eq. (3.22).*

4

Covariant Quantization

Let us recall that the closed bosonic string is described by a free bosonic field theory on a cylinder subject to additional constraints. These were formulated in eq. (2.36) and must be enforced in order to find physical states of the bosonic string. We shall discuss now how the constraints are implemented in the quantization of closed string theory.

4.1 Quantization with Constraints

There are generally two ways to quantize a theory with constraints. One is to solve the constraints in the classical theory and then to quantize the reduced system. The other is to quantize the unconstrained system and then to implement the constraints in the quantum theory through a proper selection criterion among the states. These two different ways are shown schematically in Figure 4.1.

In the case of first class constraints we can be a bit more specific. Let us begin with the left upper corner of Figure 4.1. The basic object here is the phase space $\mathscr{B}$ of the unconstrained system. It comes equipped with a POISSON structure $\{\cdot;\cdot\}_{\text{P.B.}}$. Furthermore, we have a set of constraints, which we denote by $l_\alpha, \alpha = 1, \ldots, N$.

$$
\begin{array}{ccc}
(\mathscr{B}, \{\cdot;\cdot\}_{\text{P.B.}}, l_\alpha) & \xrightarrow{\text{quantization}} & (\mathscr{H}_{\mathscr{B}}, L_\alpha) \\
{\scriptstyle \textcircled{1}}\Big\downarrow & & \Big\downarrow{\scriptstyle \textcircled{2}} \\
\left(\mathscr{A}, \{\cdot;\cdot\}_{\text{P.B.}}^{\mathscr{A}}\right) & \xrightarrow[\text{quantization}]{} & \mathscr{H}_{\mathscr{A}}
\end{array}
$$

Figure 4.1 Two quantization procedures for classical systems with first class constraints. Both paths are expected to lead to the same quantum theory in the lower right corner.

These are functions on the phase space $\mathscr{B}$ that we want to set to zero. We require the constraints l_α to be of first class, which means that the POISSON brackets should close on the set of constraints:

$$\{l_\alpha; l_\beta\}_{\text{P.B.}} = ic_{\alpha\beta}{}^\gamma l_\gamma.$$

Imposing the constraints implies that $l_\alpha = 0$. This condition selects a submanifold in the phase space $\mathscr{B}$. If dim $\mathscr{B} = 2d$ denotes the dimension of the phase space and the l_α are N independent constraints, then the constraint submanifold has dimension $2d - N$. But this is not yet the phase space of the constrained system. In fact, our first class constraints generate symmetry transformations on the constraint submanifold, and points that are related by these symmetries describe the same state of the classical theory. Hence, the phase space $\mathscr{A}$ of the constrained classical theory parametrizes orbits of the constraint submanifold (see Fig. 4.2). Since the symmetry has N generators, generic orbits are of dimension N, and consequently, the phase space of the reduced system has dimension dim $\mathscr{A} = 2d - 2N$. For first class constraints, the POISSON bracket on the phase space $\mathscr{A}$ is naturally inherited from the POISSON structure of $\mathscr{B}$. We shall denote it by the same symbol $\{\cdot; \cdot\}_{\text{P.B.}}$. Once the reduced phase space $\mathscr{A}$ and its POISSON bracket $\{\cdot; \cdot\}_{\text{P.B.}}$ are known, we can proceed to the quantization. Wave functions in the quantum theory depend on either coordinates of momenta, i.e. on $d - N$ variables.

Alternatively, we could have the idea to quantize the unconstrained system. The result of such a quantization is a state space $\mathscr{H}_\mathscr{B}$, which we can think of as the space of function in dim $\mathscr{B}/2 = d$ variables. For the latter we can select, e.g., the coordinates or the momenta of the classical system, depending on whether we develop a coordinate space or momentum space picture of the quantum system. As a result of the quantization, functions on the classical phase space get promoted to differential operators on the state space. This applies in particular to our constraints l_α. Let us denote the corresponding operators by L_α. Once $\mathscr{H}_\mathscr{B}$ and L_α are constructed, we have successfully quantized our system with constraints. Next we need to implement the constraints, i.e., we have to descend to the quantum theory of the constrained theory. Roughly speaking, the idea is to select states in $\mathscr{H}_\mathscr{B}$ as admissible wave functions of the constrained system whenever they are annihilated by the constraint operators L_α, i.e., we define

$$\mathscr{H}_\mathscr{A} := \{|\psi\rangle \in \mathscr{H}_\mathscr{B} \,|\, L_\alpha |\psi\rangle = 0\} . \tag{4.1}$$

These N differential equations remove N of the variables and hence the space of wave functions consists of functions in $d - N$ coordinates. This is more or less what is expected. Of course, one would hope that the two different ways to quantize the constrained system finally lead to the same results.

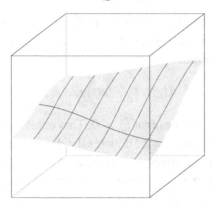

Figure 4.2 The phase space $\mathscr{B}$ showing the constraint submanifold. It is foliated by the orbits the constraints generate. The reduced phase space $\mathscr{A}$ intersects these orbits.

In the context of string theory, we have seen the left upper corner of Figure 4.1 in Chapter 2. What we want to get to is the right lower corner. Both ways of getting there have been developed. In the context of string theory, a successful implementation of the first way is known as light-cone quantization. Here we shall start by treating the string in the second approach, i.e., we will first describe the so-called covariant quantization. Note that in Chapter 3 we have quantized the bosonic free field theory, and hence we have already completed the quantization of our system with constraints. What remains to be done is to implement the constraints.

Before we address this, let us briefly look at the free massless particle. In this case, we start from a 2-dimensional phase space $\mathscr{B}$. Its coordinate functions are the D position variables x^μ and their conjugate momenta p^μ. The POISSON structure is the canonical one:

$$\{p^\mu; x^\mu\}_{\text{P.B.}} = \eta^{\mu\nu} . \tag{4.2}$$

As we discussed before, the system has one constraint $l = p^2 = 0$. Let us now quantize this system and then impose the constraint after quantization as a condition on the admissible wave functions. After quantization we obtain the state space of the unconstrained system consisting of all functions, e.g., on the position space:

$$\mathscr{H}_\mathscr{B} = \{\psi(x^\mu)\} . \tag{4.3}$$

In this representation, the momentum operators p^μ become derivatives with respect to position variables x^μ, and hence the constraint equation reads

$$L\psi = p^2\psi = -\eta^{\mu\nu}\partial_\mu\partial_\nu\psi = 0 . \tag{4.4}$$

The space of solutions, i.e., our HILBERT space $\mathcal{H}_{\mathscr{A}}$, contains all wave functions obeying the massless KLEIN-GORDON-equation:

$$\mathcal{H}_{\mathscr{A}} = \left\{ \psi(x^\mu) \mid \eta^{\mu\nu} \partial_\mu \partial_\nu \psi(x^\mu) = 0 \right\}. \tag{4.5}$$

The solutions are parametrized by the initial conditions, i.e. by functions on a space-like surface. In this way, imposing the constraint has reduced the space of states from functions on the D-dimensional MINKOWSKI space to functions on a $(D-1)$-dimensional CAUCHY surface. Alternatively, we could also work in the momentum representation where the constraint $L = p^2$ is a multiplication operator. The constraint equation $p^2 \psi(p) = 0$ is solved by functions that are supported on the light-cone, i.e., once again our state space is built from functions of $D-1$ variables.

4.2 Wave Functions of the Closed Bosonic String

We are now well prepared to tackle the covariant quantization of the closed bosonic string. As we explained earlier, the closed bosonic string is described by a free bosonic field theory on a cylinder obeying some constraints. Hence, our quantum reduction procedure begins with the state space of closed bosonic fields in D dimensions:

$$\mathcal{H}_{\mathscr{B}} = \mathcal{H}_D^{\text{CBF}}. \tag{4.6}$$

We also studied in some detail the constraints of the system, first on the classical level and then in the quantum theory. Thereby, we constructed a set of operators L_n and $\bar{L}_n$ that can be considered the quantum mechanical analogues of the classical functions l_n and $\bar{l}_n$. The only issue in the construction was related to the choice of a normal ordering prescription that affected the particular form of L_0. In this sense, we cannot be sure whether L_0 is really the correct quantum version $\hat{l}_0$ of the constraint function l_0. In fact, $\hat{l}_0$ may differ from L_0 by some finite shift, i.e., $\hat{l}_0 = L_0 - a$. After these comments we are now ready to construct our candidate state space of the bosonic string:

$$\mathcal{H}'_{\mathscr{A}} = \mathcal{H}'_{D,a} = \left\{ |\psi\rangle \in \mathcal{H}_{\mathscr{B}} \; \middle| \; \begin{array}{ll} L_n |\psi\rangle = \bar{L}_n |\psi\rangle = 0 & \text{for } n > 0 \\ L_0 |\psi\rangle = \bar{L}_0 |\psi\rangle = a |\psi\rangle & \end{array} \right\}. \tag{4.7}$$

As we remarked above, the constant a accounts for normal ordering ambiguities in the quantization of l_0. Later we shall see that $a = (D-2)/24$, but we will keep it as a free parameter for now. Another comment concerns our implementation of the condition $L_n = 0$. We have restricted it to $n > 0$ and thereby impose $L_n = 0$ only in the weak sense, i.e., such that all matrix elements of the L_n vanish:

$$\langle \psi \,|\, L_n \,|\, \psi' \rangle \;=\; 0 \qquad for \quad |\psi\rangle, |\psi'\rangle \in \mathcal{H}'_{\mathcal{A}} . \tag{4.8}$$

If we would impose $L_n \,|\, \psi \rangle \;=\; \bar{L}_n \,|\, \psi \rangle \;=\; 0$ for all $n \in \mathbb{Z}$, no states would be left in the state space because of the non-vanishing central charge term in the commutator of L_n and L_{-n}. Vanishing of the matrix elements is considerably weaker but sufficiently strong to implement our constraints.

Our introductory remarks suggest that we are done with the quantization, but there remains a small complication that we need to investigate in more detail. Before we get there, let us recall that the space $\mathcal{H}_{\mathcal{B}}$ contains plenty of negative norm states. In order for elements of $\mathcal{H}'_{\mathcal{A}}$ to have an interpretation as a wave function of the bosonic string, it is necessary that somehow imposing the constraints has removed all those negative norm states. This is indeed the case, at least for a particular choice of the dimension D and the constant a.

According to the so-called *No Ghost Theorem of* Brower-Goddard-Thorn [14, 26] the constrained state space $\mathcal{H}'_{\mathcal{A}} = \mathcal{H}'_{D=26, a=1}$ of the closed bosonic string does not contain any negative norm states for $D = 26$ and $a = 1$.

The proof is certainly beyond our scope here, but at least we would like to understand how our requirement of removing negative norm states can lead to conditions on a and D. As an example, let us look at the following states:

$$|\xi, k\rangle \;=\; \xi_\mu a^\mu_{-1} \,|\, k \rangle \ . \tag{4.9}$$

We know from the last chapter that its norm is given by

$$\|\,|\xi, k\rangle\,\|^2 \;\sim\; \xi_\mu \xi^\mu, \tag{4.10}$$

which is negative for time-like polarizations ξ. So, somehow, our constraints must eliminate all such unwanted polarizations. All constraints of the form $L_n \,|\, \xi, k\rangle = 0$ for $n \geq 2$ are trivially satisfied. Hence it remains to study those of L_1 and L_0. From the constraint involving L_1 we deduce

$$L_1 \,|\, \xi, k\rangle \;=\; 0 \qquad \Rightarrow \qquad k_\mu \xi^\mu = 0, \tag{4.11}$$

and the L_0-constraint gives

$$L_0 \,|\, \xi, k\rangle \;=\; a \,|\, \xi, k\rangle \qquad \Rightarrow \qquad 1 + \frac{\alpha'}{4} k^2 = a \ . \tag{4.12}$$

The first condition tells us that physical states of the bosonic string must be polarized such that $k_\mu \xi^\mu = 0$. In order to exclude the unwanted time-like polarizations, we must select time-like or light-like momenta k. The latter are selected if we impose the second condition with $a \leq 1$. Hence, in order to get rid of the negative norm states, we have to take $a \leq 1$. This does not yet tell us to pick $a = 1$ and also has not provided any help with the conditions on the dimension D (see, however, problem 10), but it may suffice to understand the general mechanism by which such conditions can arise. We have seen that for $a = 1$ the negative norm states on the first excited level are indeed removed.

Our example has also taught us another lesson: that $\mathcal{H}'_{D=26,a=1}$ still contains states of vanishing norm. The latter do not allow for a sensible interpretation as wave functions of the bosonic string. But it is consistent to simply set them to zero. Thereby we now arrive at the following improved definition for the state space of the closed bosonic string:

$$\mathcal{H}^{\text{CBS}}_{D=26} = \mathcal{H}'_{D=26,a=1} \big/ \mathcal{N} , \tag{4.13}$$
$$\text{where} \quad \mathcal{N} = \left\{ |\psi\rangle \in \mathcal{H}'_{D=26,a=1} \,|\, \langle \psi | \psi \rangle = 0 \right\} . \tag{4.14}$$

By construction, the space $\mathcal{H}^{\text{CBS}}_{D=26}$ comes equipped with a positive definite bilinear form, i.e., it is the HILBERT space of wave functions for the *closed bosonic string* (CBS).

4.3 Mass Spectrum of the Closed Bosonic String

As we explained in the introduction, closed strings give rise to an infinite set of vibrational modes that we can think of as an infinite set of particles. In the last section we have constructed all the wave functions of the bosonic string, and hence we should be curious to find out what kind of particles they describe. The simplest observable quantity of a particle is its mass $M^2 = -p^\mu p_\mu$. Since the momentum operators p^μ commute with our VIRASORO constraints, they descend from the state space of the field theory to the state space of string theory. Hence, M^2 is a well-defined operator on $\mathcal{H}^{\text{CBS}}_{D=26}$, and we would like to find its spectrum. In order to do so, we consider the VIRASORO generator L_0 and rewrite it in the form

$$L_0 = \frac{1}{2} \sum_{n \in \mathbb{Z}} : a^\mu_{-n} a^\nu_n : \eta_{\mu\nu}$$
$$= \frac{1}{2} a^\mu_0 a^\nu_0 \, \eta_{\mu\nu} + \sum_{n=1}^{\infty} a^\mu_{-n} a^\nu_n \, \eta_{\mu\nu} \tag{4.15}$$
$$= \frac{\alpha'}{4} p^2 + \sum_{n=1}^{\infty} n \, N_n .$$

Here, the number operators N_n count the number of creation operators a_{-n}. In an analogous formula for $\bar{L}_0$, the operator $\bar{N}_n$ counts the number of creation operators

$\bar{a}_{-n}$. Inserting $p^2 = -M^2$ and using that $L_0 = 1$ on all wave functions of the bosonic string, we obtain the following mass formula on the HILBERT space $\mathcal{H}^{CBS}_{D=26}$:

$$M^2 = -\frac{4}{\alpha'} + \frac{4}{\alpha'} \sum_{n=1}^{\infty} n N_n = -\frac{4}{\alpha'} + \frac{4}{\alpha'} \sum_{n=1}^{\infty} n \bar{N}_n . \tag{4.16}$$

Our first, somewhat disappointing, observation is that the lowest possible eigenvalue for M^2 is $M^2 = -4/\alpha'$, i.e., it is negative. This means that the spectrum of the closed bosonic string contains a tachyonic excitation, a fact that will keep us busy for another few chapters. Beyond the tachyon, things start to look better. All excited states with at least one $N_n = \bar{N}_n \neq 0$ possess non-negative mass. Massless states are possible for $N_1 = \bar{N}_1 = 1$ and $N_n = 0 = \bar{N}_n$ for $n \geq 2$.

Let us analyze the exact particle content of the two lowest mass levels in some more detail. At the lowest level with no oscillators, the most general state of the closed bosonic field theory is a superposition of the eigenfunctions $|k\rangle$ of the momentum operator p^μ,

$$|\phi\rangle = \int d^D k \, |\phi; k\rangle = \int d^D k \, \hat{\phi}(k) \, |k\rangle ,$$

where $\hat{\phi}(k)$ stands for the amplitude of the momentum k component. Imposing our physical state constraints, we obtain a number of conditions on the amplitude $\hat{\phi}(k)$. The L_0 constraint gives

$$L_0 |\phi\rangle = \int d^D k \, \frac{\alpha'}{4} k^2 \hat{\phi}(k) \, |k\rangle = \int d^D k \, \hat{\phi}(k) \, |k\rangle \tag{4.17}$$

so that the amplitude $\hat{\phi}(k)$ must satisfy

$$k^2 \hat{\phi}(k) = \frac{4}{\alpha'} \hat{\phi}(k) . \tag{4.18}$$

Hence $\hat{\phi}(k)$ is non-zero only on the hyperboloid for mass $M^2 = -4/\alpha'$. Defining a field ϕ through FOURIER transform of the amplitude $\hat{\phi}(k)$ we obtain

$$\Box \phi(x) + M^2 \phi(x) = 0 ,$$

i.e. ϕ satisfies the massive KLEIN-GORDON equation. Thereby, we have shown that the lowest mass level in the spectrum of closed bosonic strings gives rise to a single tachyonic scalar field.

The next mass level has one left- and one right-moving oscillator excited. These massless states must be obtained from a superposition of the following eigenstates of the momentum operator:

$$|\Omega; k\rangle = \Omega_{\mu\nu}(k) \, a^\mu_{-1} \bar{a}^\nu_{-1} \, |k\rangle , \tag{4.19}$$

where $\Omega(k)$ is a matrix of amplitudes for the eigenstates of p with momentum k and $N_1 = 1 = \bar{N}_1$. The non-trivial constraints read

$$L_1 \mid \Omega; k\rangle = 0 \qquad\qquad \Leftrightarrow k^\mu \Omega_{\mu\nu}(k) = 0, \qquad (4.20a)$$

$$\bar{L}_1 \mid \Omega; k\rangle = 0 \qquad\qquad \Leftrightarrow \Omega_{\mu\nu}(k)k^\nu = 0, \qquad (4.20b)$$

$$L_0 \mid \Omega; k\rangle = \bar{L}_0 \mid \Omega, k\rangle = \mid \Omega, k\rangle \qquad \Leftrightarrow k^2 = 0. \qquad (4.20c)$$

Note that the space of allowed polarizations $\Omega_{\mu\nu}(k)$ depends on the choice of the light-like momentum. Since all light-like momenta are related by SO(1,25) transformations, it suffices to analyze the space of solutions to eqs. (4.20) for a light-like momentum k of the form $k = (1, 1, 0, \ldots, 0)$. Once this choice is adopted, the two transversality conditions for the polarization tensor Ω read

$$\Omega_{0\nu} + \Omega_{1\nu} = 0, \qquad (4.21a)$$

$$\Omega_{\mu0} + \Omega_{\mu1} = 0. \qquad (4.21b)$$

Let us now compute the norm of the states $\mid \Omega, k\rangle$,

$$
\begin{aligned}
\langle\Omega, k \mid \Omega, k\rangle &= \Omega_{\mu\nu}\Omega^{\mu\nu} \\
&= \Omega_{0\nu}\Omega^{0\nu} + \Omega_{1\nu}\Omega^{1\nu} + \Omega_{\alpha0}\Omega^{\alpha0} + \Omega_{\alpha1}\Omega^{\alpha1} + \Omega_{\alpha\beta}\Omega^{\alpha\beta} \quad (4.22) \\
&= \left(\Omega_{\alpha\beta}\right)^2 \geq 0,
\end{aligned}
$$

where $\alpha, \beta \in \{2, 3, \ldots, 25\}$. We have used the transversality conditions to drop the first four terms in the second line of the above computation. The result tells us that for given momentum k the space of non-zero wave functions with $M^2 = 0$ coincides with the space of 24×24 matrices. The wave functions carry an action of SO(24) $\subset$ SO(1, 25), the subgroup that leaves our momentum vector $k = (1, 1, 0, \ldots, 0)$ invariant. This group is often referred to as the *little group* for massless states. On the 24×24 matrices that describe polarizations of massless states, the little group SO(24) acts by simple conjugation.

We are interested to find the particle multiplets that are described by the massless states of the closed bosonic string. By definition, a massless multiplet corresponds to an irreducible representation of the massless little group on the space of polarizations. Hence, our task is to decompose the space of 24×24 matrices $(\Omega_{\alpha\beta})$ into irreducible subrepresentations under the adjoint action of SO(24). With a bit of support from mathematics, this is not that hard too accomplish, and we discover three multiplets[1]:

---

[1] The mathematics of this decomposition is the same as for the decomposition of 3×3 matrices under the action of SO(3). In this case, the irreducible multiplets arise from the composition of two vector (spin 1) representations:

$$[J = 1] \times [J = 1] = [J = 2] \oplus [J = 1] \oplus [J = 0].$$

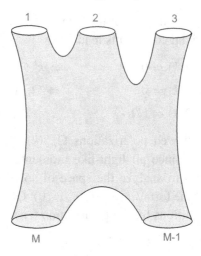

Figure 4.3 A tree-level closed scattering diagram can be decomposed into $M - 2$ elementary vertices. Each of these vertices contributes one factor g_s while each external state absorbs a factor g_s, giving a total g_s^{M-2}.

- a symmetric traceless part with $\Omega_{\alpha\beta} = \Omega_{\beta\alpha}$ and $\Omega_\alpha{}^\alpha = 0$, representing a spin 2 particle, the graviton; see, e.g., [70],
- an antisymmetric part with $\Omega_{\alpha\beta} = -\Omega_{\beta\alpha}$ representing a 2-form-field[2] known as KALB-RAMOND-field (or B-field),
- a part proportional to the identity $\Omega_{\alpha\beta} = \xi \mathbb{1}_{24}$, representing a scalar particle, the dilaton.

This concludes our investigation of the massless states of a closed bosonic string.

4.4 4-Point Tachyon Amplitude

At this point we have collected enough techniques to compute scattering amplitudes of some simple closed string modes. Our aim here is to derive the VIRASORO-SHAPIRO amplitude for the scattering amplitude of four tachyonic modes of the closed string. Before we get there, however, we need to discuss the general prescription for the computation of scattering amplitudes.

A string scattering amplitude assigns a number to a set of states $\psi_i \in \mathcal{H}_{D=26}^{\text{CBS}}$ of the closed bosonic string. We will soon see how this is achieved. Let us recall that any closed string scattering amplitude with M external legs and without holes can be cut into $M - 2$ simple vertices with three legs. The latter are weighted with a coupling g_s, and hence the M-point vertex is proportional to g_s^{M-2}. In order to

---

[2] In three dimensions (and only there) a 2-form-field would be equivalent to a vector field.

weight all tree amplitudes with the same power of g_s, we absorb one factor of g_s into each leg. Hence, closed string tree amplitudes are proportional to g_s^{-2}. The unique string tree-level FEYNMAN graph with M external legs is depicted in Figure 4.3.

We shall now explain how the computation of string scattering amplitudes is reduced to the computation of correlation functions in the corresponding world-sheet (free) field theory. The basic idea is simple: We have stated very generally that for any string wave function ψ_i there exists a unique field $V_{\psi_i} = V_i$ such that $V_i(0) | 0 \rangle = | \psi_i \rangle$. For wave functions of closed strings, these fields are inserted at some point $(z, \bar{z})$ of the complex plane. The insertion point does not correspond to any target space data, and it needs to be removed somehow. We do this simply by integrating over all the M insertion points $(z_i, \bar{z}_i)$. These remarks motivate the following formal definition of the scattering amplitude:

$$\mathcal{A}(\psi_1, \ldots, \psi_M) \sim g_s^{-2} \langle \mathcal{O}_{\psi_1} \cdots \mathcal{O}_{\psi_M} \rangle_{\text{FT}}, \tag{4.23}$$

where $\mathcal{O}_\psi$ are integrated vertex operators:

$$\mathcal{O}_\psi = \int_{\mathbb{C}} \mathrm{d}^2 z \, V_\psi(z) . \tag{4.24}$$

If we are interested in the amplitude of closed string tachyons, for example, the external states are given by $| k_i \rangle$ with $k_i^2 = 1/\alpha$, and the corresponding fields are our usual exponentials V_k. The definition (4.23) is morally correct but suffers from a little defect that still needs to be cured before we have a well-defined construction of scattering amplitudes.

The suggested formula turns out to give infinite results for a very simple reason. The correlation functions in our free field theory have some symmetries. In particular they do not change if we, e.g., translate all insertion points $(z_i, \bar{z}_i)$ by the same complex constant b. When we integrate over insertion points, we also integrate over all those simple translations, and if the integrand is always the same, then the integral is infinite. Translation invariance is actually not the only such symmetry of the field theory correlators. Closer inspection shows that they are covariant under all rational transformations of the complex plane, i.e., under all the maps

$$u \mapsto \frac{au + b}{cu + d} \quad \text{where} \quad ad - bc = 1;$$

see problem 12. Here all parameters a, b, c, d are taken to be complex, and the rational transformations we described provide an action of the non-compact group $SL(2, \mathbb{C})$. Note that such transformations include the simple translations $z \mapsto z + b$ we talked about before. What we conclude from all this is that the amplitude $\mathcal{A}$ we defined above is proportional to the volume of $SL(2, \mathbb{C})$. In order to get something finite we need to divide by this volume. Since the 3-parameter group of rational transformations can be used to put three insertions points into $z_1 = \infty, z_2 = 0$, $z_3 = 1$, the correct prescription is to fix these three insertions and then to integrate

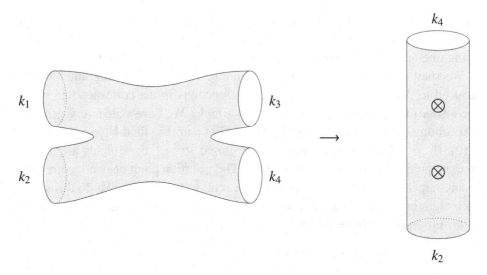

Figure 4.4 The string FEYNMAN graph for the 4 point function is mapped to a cylinder with two operator insertions.

only over the remaining $M - 3$ insertion points $z_4, \ldots, z_M$. Therefore, the prescription for the computation of amplitudes becomes

$$
\mathcal{A}(\psi_1, \ldots, \psi_M) = g_s^{-2} \int d^2 z_4 \cdots d^2 z_M
$$

$$
\langle V_1(\infty) V_2(0) V_3(1) V(z_4, \bar{z}_4) \cdots V_M(z_M, \bar{z}_M) \rangle . \quad (4.25)
$$

Let us recall that the correlation function of local fields in the free field theory are analytic in the complex plane. Therefore, we do not need to order the arguments such that $|z_i| > |z_{i+1}|$. As we have learned before, $V_k(0)|0\rangle = |k\rangle$. Similarly, one can argue that $\langle 0|V_k(\infty) = \langle -k|$. Let us now determine the scattering amplitude for four closed string tachyons. For each tachyon wave function $|k\rangle$ with $k^2 = \frac{4}{\alpha'}$ there exists a field operator

$$
V_k(z, \bar{z}) =: \exp(ik \cdot X(z, \bar{z})) : . \quad (4.26)
$$

The 4 point amplitude as depicted in Figure 4.4 corresponds to the cylinder with two operator insertions

$$
\mathcal{A}^{4\mathrm{T}} = \int \langle -k_1 \,|\, V_{k_3}(1) V_{k_4}(z, \bar{z}) \,|\, k_2 \rangle \, d^2 z. \quad (4.27)
$$

We will now express this amplitude in MANDELSTAM variables s and t:

$$
s = -(k_1 + k_2)^2 = -(k_3 + k_4)^2 = -k_4^2 - k_3^2 - 2k_4 \cdot k_3 = -\frac{8}{\alpha'} - 2k_4 \cdot k_3 ,
$$

$$
(4.28\mathrm{a})
$$

$$
t = -(k_1 + k_3)^2 = -(k_2 + k_4)^2 = -k_3^2 - k_4^2 - 2k_2 \cdot k_4 = -\frac{8}{\alpha'} - 2k_2 \cdot k_4.
$$

$$
(4.28\mathrm{b})
$$

Introducing $\alpha(x) = \frac{\alpha'}{2}x + 2$, we can write the amplitude as

$$
\begin{aligned}
\mathcal{A}^{4\mathrm{T}} &= \int |z|^{\alpha' k_4 \cdot k_2} |z-1|^{\alpha' k_4 \cdot k_3} \mathrm{d}^2 z \\
&= \int |z|^{\alpha'\left(-\frac{1}{2}t - \frac{4}{\alpha'}\right)} |z-1|^{\alpha'\left(-\frac{1}{2}s - \frac{4}{\alpha'}\right)} \mathrm{d}^2 z \\
&= \pi \frac{\Gamma\left(-\frac{1}{2}\alpha(s)\right) \Gamma\left(-\frac{1}{2}\alpha(t)\right) \Gamma\left(1 + \frac{1}{2}\alpha(s) + \frac{1}{2}\alpha(t)\right)}{\Gamma\left(1 + \frac{1}{2}\alpha(s)\right) \Gamma\left(1 + \frac{1}{2}\alpha(t)\right) \Gamma\left(-\frac{1}{2}\alpha(s) - \frac{1}{2}\alpha(t)\right)} \\
&\approx \frac{1}{\frac{\alpha'}{4}t + 1} + \frac{\left(1 + \frac{1}{2}\alpha(s)\right)\alpha(s)}{\frac{\alpha'}{4}t} + \cdots.
\end{aligned}
\tag{4.29}
$$

In the last line, we wrote out the leading poles, of which the first one corresponds to the scalar tachyon with $m^2 = -\frac{4}{\alpha'}$ and the second one corresponds to the massless fields, the dilaton, the B-field, and the graviton. While our prescription to compute scattering amplitudes might have appeared a little ad hoc, we have now demonstrated that it passes a very important consistency check: The excitations that propagate in the intermediate channel belong to the physical spectrum of closed bosonic string.

Exercises

Problem 10. *A 2-parameter family of states $|\Psi\rangle = |\Psi(\alpha, \beta)\rangle$ is defined by*

$$
|\Psi\rangle = \left(a^\mu_{-1} a_{-1,\mu} + \alpha a^\mu_0 a_{-2,\mu} + \beta (a^\mu_0 a_{-1,\mu})^2\right) |k\rangle,
$$

where $k^2 = -4/\alpha'$. Determine α, β such that $L_1|\Psi\rangle = 0 = L_2|\Psi\rangle$ and show that the norm of the resulting state is negative when $D > 26$.

Problem 11. *Conformal transformations $z \mapsto \gamma(z)$ of the complex plane are defined by the prescription*

$$
\gamma(z) = \frac{az+b}{cz+d} \quad \text{with} \quad ad - bc = 1.
$$

Show that the correlation functions (3.22) transform under conformal transformations as follows:

$$
\left\langle 0 \left| \prod_{i=1}^n V_{k_i}(\gamma(z_i)) \right| 0 \right\rangle = \prod_{i=1}^N \left| \frac{d\gamma(z_i)}{dz_i} \right|^{-\alpha' k_i^2} \left\langle 0 \left| \prod_{i=1}^n V_{k_i}(z_i) \right| 0 \right\rangle.
$$

5

Light-Cone Quantization

In the last two chapters we described the covariant quantization of the closed bosonic string, i.e., we quantized free bosonic field theory along with the infinite number of constraints $l_n = 0 = \bar{l}_n$ and then imposed the constraint in the quantum theory. The following chapter pursues a different route in which we solve the constraints first in the classical theory and then quantize the reduced theory. This alternative procedure is known as light-cone quantization. It will ultimately lead to the same results, but the restriction to $D = 26$ will arise from a very different mechanism. As in all the previous chapters, it is instructive to study the light-cone quantization for massless relativistic particles first, before turning to the string.

5.1 The Relativistic Particle Revisited

In the second chapter, we described the massless relativistic particle as a constrained system. The formulation departed from a $2D$-dimensional phase space $\mathscr{B} = \mathbb{R}^{2D}$ that is generated by the coordinates x^μ, p^μ with POISSON bracket

$$\{p^\mu; x^\nu\}_{\text{P.B.}} = \eta^{\mu\nu}. \tag{5.1}$$

Note that we explicitly included the pair of conjugate coordinates x^0, p^0 for the time direction. Thereby, the phase space is manifestly covariant. On the other hand, the true phase space of a relativistic particle has dimension $2D - 2$. As we have seen before, two directions are removed by the constraint

$$l = p^2 = p^\mu p_\mu = 0. \tag{5.2}$$

The function l generates a gauge symmetry on $\mathscr{B}$ via the POISSON bracket. It is given by

$$\{l; p^\mu\}_{\text{P.B.}} = 0 \qquad \{l; x^\mu\}_{\text{P.B.}} = 2p^\mu = 2\frac{d}{d\tau}X^\mu(\tau), \tag{5.3}$$

i.e., the constraint l generates shifts of the world-line time τ. Thus, we can use the gauge symmetry generated by l to set one of the coordinates $X^\mu(0) = x^\mu$ equal to zero. In other words, we can fix τ such that $\tau = 0$ when the particle moves through the surface $x^\mu = 0$. This eliminates the coordinate x^μ from the set of non-trivial phase space coordinates.

For reasons that will become clear in a moment, we shall actually set a particular linear combination of x^0 and x^1 to zero. To this end, let us introduce the so-called light-cone coordinates and momenta:

$$x^\pm = \frac{1}{\sqrt{2}}(x^0 \pm x^1), \qquad p^\pm = \frac{1}{\sqrt{2}}(p^0 \pm p^1). \tag{5.4}$$

The light-cone gauge amounts to setting x^+ (or x^-, but not both at the same time) to zero:

$$x^+ = X^+(\tau = 0) = 0. \tag{5.5}$$

Through this condition, we have picked one point on each gauge orbit that is generated by the function l. Thereby, we have removed one coordinate from the $2D$ phase space $\mathcal{B}$ we started with. But the resulting $(2D - 1)$-dimensional space still contains points that do not describe states of the constrained system. In fact, we still need to impose the constraint $l = -2p^+p^- + p^\alpha p_\alpha = 0$ itself. We can use this constraint to express the momentum p^-, i.e., the one conjugate to x^+, through the other momenta,

$$p^- = \frac{1}{2p^+} p^\alpha p_\alpha, \tag{5.6}$$

where we sum over $\alpha = 2, \ldots, D - 1$. We see here that light-cone gauge is particularly nice, since it allows us to construct the redundant momentum p^- from the other momenta without taking square roots.

The relation (5.6) means that we no longer need the momentum p^- so that our system is now described by a $(2D - 2)$-dimensional reduced phase space $\mathcal{A} = \mathbb{R}^{2(D-1)}$. The latter has the coordinates $x^-, p^+, x^\alpha, p^\alpha$ satisfying the Poisson algebra:

$$\left\{ p^+; x^- \right\}_{\text{P.B.}} = -1, \qquad \left\{ p^\alpha; x^\beta \right\}_{\text{P.B.}} = \delta^{\alpha\beta}. \tag{5.7}$$

To construct classical observables of the reduced system we must select all functions $A \in Fun(\mathcal{B})$ that POISSON commute with the constraint l, i.e. $\{l, A\} = 0$. From those we obtain the observables of the reduced system by restriction to the constrained surface $l = 0$. In practice, given any such function A of the $2D$ coordinates

x^μ and momenta p^μ, we descend to a function A on the reduced phase space by setting $x^+ = 0$ and $p^- = p^\alpha p_\alpha / 2p^+$.[1]

The relativistic particle has some particularly important observables that we would like to discuss in some more detail, namely the generators $M^{\mu\nu}$ and P^μ of the POINCARÉ algebra:

$$M^{\mu\nu} = x^\mu p^\nu - x^\nu p^\mu, \quad P^\mu = p^\mu. \tag{5.8}$$

These functions generate infinitesimal LORENTZ transformations and translations through their POISSON commutator with other functions on the phase space. Since they POISSON commute with the constraint l, i.e. $\{l, M^{\mu\nu}\} = 0 = \{l, P^\mu\}$, they represent observables of the reduced system. Moreover, on the full phase space $\mathscr{B}$, they obey the relations of a POINCARÉ-POISSON algebra:

$$\{M^{\mu\nu}; M^{\rho\sigma}\}_{\text{P.B.}} = \eta^{\nu\rho} M^{\mu\sigma} - \eta^{\nu\sigma} M^{\mu\rho} + \eta^{\mu\sigma} M^{\nu\rho} - \eta^{\mu\rho} M^{\nu\sigma},$$
$$\{M^{\mu\nu}; p^\rho\}_{\text{P.B.}} = \eta^{\nu\rho} p^\mu - \eta^{\mu\rho} p^\nu, \tag{5.9}$$
$$\{p^\mu; p^\nu\}_{\text{P.B.}} = 0.$$

But in passing to a description of the reduced system, we broke POINCARÉ symmetry through our light-cone gauge condition. So, one may wonder what has happened to the symmetry generators and, in particular, whether they still satisfy the same algebraic relations after expressing them in terms of x^-, p^+, x^α, and p^α. Of course, things just work out fine. We will not discuss that in full generality, but look at one example to get some intuition on how to verify the POINCARÉ symmetry of the reduced system.

To begin, let us pick the generators $M^{-\alpha}$ and see what they look like when written as functions on the reduced phase space $\mathscr{A}$. Recall that we are instructed to set $x^+ = 0$ and $p^- = p^\alpha p_\alpha / 2p^+$ so that $M^{-\alpha}$ becomes

$$M^{-\alpha} = x^- p^\alpha - x^\alpha \frac{p_\beta p^\beta}{2p^+}. \tag{5.10}$$

On the phase space $\mathscr{B}$, the POISSON bracket of the functions $M^{-\alpha}$ with the momenta p^α is given by $\{M^{-\alpha}; p^\alpha\}_{\text{P.B.}} = p^-$. Let us check that this is still true when we use the expression (5.10) along with the POISSON algebra (5.7) on $\mathscr{A}$:

$$\{M^{-\alpha}; p^\alpha\}_{\text{P.B.}} = -\left\{\frac{x^\alpha p^\beta p_\beta}{2p^+}; p^\alpha\right\}_{\text{P.B.}} = p^-. \tag{5.11}$$

---

[1] Note that the functions x^μ do not commute with the constraint, and hence, by definition, they do not give rise to observables. On the other hand, the quantities $x^\mu - p^\mu x^-/p^-$ do commute with the constraint, and upon setting $x^- = 0$ they agree with x^μ.

Similarly, one can verify all the other POISSON brackets between generators of the POINCARÉ symmetry of the massless relativistic particles.

The quantization of the constrained system is straightforward once we adopt its description through the reduced phase space $\mathscr{A}$. The HILBERT space of states $\mathscr{H}_{\mathscr{A}} = L^2(\mathbb{R}^{D-1})$ in the momentum representation consists of all functions in the $D - 1$ variables p^+ and p^α. It is spanned by eigenstates $|k^+, k^\alpha\rangle$ of the momentum operators. On these functions, the coordinates act as differential operators:

$$x^\alpha \to \mathrm{i}\frac{\partial}{\partial p^\alpha}, \qquad x^- \to -\mathrm{i}\frac{\partial}{\partial p^+}. \tag{5.12}$$

All observables of the classical system give rise to operators on the state space. Writing these operators requires us to fix the usual ordering ambiguities. Here, we shall assume a symmetric ordering prescription. With this convention, the operator $M^{-\alpha}$ takes the following form:

$$M^{-\alpha} = x^- p^\alpha - \frac{1}{2}(p^- x^\alpha + x^\alpha p^-). \tag{5.13}$$

All other generators of the POINCARÉ symmetry can be constructed similarly. The commutation relations between the resulting operators reproduce the defining relations of the POINCARÉ algebra. As an example one may easily check that

$$\left[M^{-\alpha}, M^{-\beta}\right] = 0.$$

Through computations of this type one can establish that the space of one-particle wave functions of a relativistic particle carries a representation of the POINCARÉ algebra. The outcome is certainly not surprising.

5.2 The Relativistic String Revisited

We are now prepared to move on to the bosonic string. Our aim is again to quantize the system only after solving (almost) all constraints. We will follow the particle case as closely as possible. Recall that the constraints for the closed bosonic strings were obtained from the LAURENT coefficients l_n and $\bar{l}_n$ of the stress energy tensor. We have also discussed that the modes generate diffeomorphisms on the space of field configurations. The idea is to use these diffeomorphisms in order to achieve that

$$X^+(\sigma, 0) = x^+ + \mathrm{i}\sqrt{\frac{\alpha'}{2}} \sum_{n\neq 0} \left(\frac{a_n^+}{n}e^{-\mathrm{i}n\sigma} + \frac{\bar{a}_n^+}{n}e^{\mathrm{i}n\sigma}\right) \equiv 0. \tag{5.14}$$

For the time-dependence of X^+ the choice implies that $X^\alpha(\sigma,\tau) = p^+\tau$. In terms of the usual oscillators, our gauge condition reads

$$x^+ = 0, \quad a_n^+ = 0, \quad \bar{a}_n^+ = 0 \tag{5.15}$$

for all $n \neq 0$. Note that we exploited the gauge symmetries generated by l_n and $\bar{l}_n$ to trivialize a_n^+ and $\bar{a}_n^+$. In the space of zero-modes, we only set $x^+ = 0$ with the help of $l_0 + \bar{l}_0$. Consequently, the gauge symmetry $l_0 - \bar{l}_0$ is left. We could use it to remove one of the other zero modes of the theory. But it is more convenient to keep $l_0 - \bar{l}_0$ as a constraint that we impose after quantizing the system.

The reduced phase space $\mathscr{A}'$ of our reduced theory is the infinite dimensional space parametrized by $x^-, p^+, x^\alpha, p^\alpha, a_n^\alpha, \bar{a}_n^\alpha$, with the canonical POISSON bracket as well as one constraint, $l_0 = \bar{l}_0$. As in the case of the relativistic particle, we can express the coordinates p^-, a_n^-, and $\bar{a}_n^-$ through coordinates of the reduced phase space by solving the constraints:

$$a_n^- = \sqrt{\frac{2}{\alpha'}}\frac{1}{2p^+}\sum_{\alpha=2}^{D-1}\sum_{m\in\mathbb{Z}}a_{n-m}^\alpha a_m^\alpha = \sqrt{\frac{2}{\alpha'}}\frac{l_n^\perp}{p^+}, \tag{5.16}$$

$$\bar{a}_n^- = \sqrt{\frac{2}{\alpha'}}\frac{1}{2p^+}\sum_{\alpha=2}^{D-1}\sum_{m\in\mathbb{Z}}\bar{a}_{n-m}^\alpha \bar{a}_m^\alpha = \sqrt{\frac{2}{\alpha'}}\frac{\bar{l}_n^\perp}{p^+}, \tag{5.17}$$

$$p^- = \frac{1}{\alpha'p^+}(l_0^\perp + \bar{l}_0^\perp). \tag{5.18}$$

Here, we have introduced the objects $l_n^\perp$ and $\bar{l}_n^\perp$ through

$$l_n^\perp = \frac{1}{2}\sum_m :a_{n-m}^\alpha a_{m;\alpha}:, \quad \bar{l}_n^\perp = \frac{1}{2}\sum_m :\bar{a}_{n-m}^\alpha \bar{a}_{m;\alpha}:,$$

where the summation in the index $\alpha = 2, \ldots, D-1$, extends over $D-2$ transverse directions. Our expression for a_n^-, for example, is obtained from the constraint $l_n = 0$ for $n \neq 0$:

$$l_n = \frac{1}{2}\sum_{m\in\mathbb{Z}}(-2a_{n-m}^+ a_m^- + a_{n-m}^\alpha a_m^\alpha) = -p^+\sqrt{\frac{\alpha'}{2}}a_n^- + l_n^\perp = 0. \tag{5.19}$$

In the second equality for l_n we have used that $a_m^+ = 0$, for $m \neq 0$ and we inserted $a_0^+ = p^+\sqrt{\alpha'/2}$. Similarly, the constraint $l_0 + \bar{l}_0 = 0$ gives

$$l_0 + \bar{l}_0 = -p^+\alpha'p^- + l_0^\perp + \bar{l}_0^\perp = 0, \tag{5.20}$$

from which our expression for p^- follows. Let us note in passing that the functions $l_n^\perp$ and $\bar{l}_n^\perp$ possess the same algebraic properties as l_n and $\bar{l}_n$, i.e., they obey the relations of a WITT-POISSON algebra. One may consider this appearance of the WITT algebra accidental. The generators $l_n^\perp$ do not generate a gauge symmetry of the classical theory.

5.3 Light-Cone Quantization of the Bosonic String

The quantization of the light-cone gauged closed bosonic string is fairly easy now. To begin, we construct the state $\mathcal{H}'_{\mathcal{A}}$ by

$$\mathcal{H}'_{\mathcal{A}} = \int d^{D-2}k^\perp dk^+ \, \bar{\mathcal{H}}^\perp_{k^\perp,k^+} \otimes \mathcal{H}^\perp_{k^\perp,k^+}.$$

The spaces $\mathcal{H}_{k^\perp,k^+}$ are constructed out of the ground states $|k^\perp, k^+\rangle$ by application of the transverse creation operators a^α_{-n} with $n > 0$ and $\alpha = 2, \ldots, D-1$. We have placed a $'$ on the symbol $\mathcal{H}$ to remind us of the remaining constraint $L_0 = \bar{L}_0$. The space of states of the closed bosonic string finally reads

$$\mathcal{H}_{\mathcal{A}} = \{|\psi\rangle \in \mathcal{H}'_{\mathcal{A}} \,|\, L_0^\perp |\psi\rangle = \bar{L}_0^\perp |\psi\rangle\}. \tag{5.21}$$

Here, we have used that $(L_0 - \bar{L}_0)|\psi\rangle = (L_0^\perp - \bar{L}_0^\perp)|\psi\rangle$ for all states $|\psi\rangle \in \mathcal{H}'_{\mathcal{A}}$. The constraint on states is also known as *level matching* condition.

As in the covariant construction, the space of wave functions comes equipped with the mass operator. On $\mathcal{H}'_{\mathcal{A}}$, the square of the mass operator reads

$$M^2 = 2p^+p^- - p^\alpha p_\alpha = \frac{4}{\alpha'}\left(L_0^\perp - \frac{1}{2}\sum_{\alpha=2}^{D-1} a_0^\alpha a_0^\alpha - a\right) = \frac{4}{\alpha'}(N^\perp - a) \tag{5.22}$$

$$= \frac{4}{\alpha'}\left(\bar{L}_0^\perp - \frac{1}{2}\sum_{\alpha=2}^{D-1} a_0^\alpha a_0^\alpha - a\right) = \frac{4}{\alpha'}(\bar{N}^\perp - a), \tag{5.23}$$

where a is the constant that arises because of the normal ordering in the quantization of the theory. As we have discussed in the previous chapter, one should think of $L_0^\perp - a$ as the proper quantization of the function $l_0^\perp$. In the present context, there is some highly suggestive computation that supports the value $a = 1$ we chose in the covariant quantization:

$$\frac{\alpha'}{2}p^+p^- = \frac{1}{2}\sum_{\substack{m\in\mathbb{Z} \\ \alpha=2,\ldots,D-1}} a^\alpha_{-m}a^\alpha_m = \frac{1}{2}\sum_{\alpha=2}^{D-1}\left(a_0^\alpha a_0^\alpha + 2\sum_{m=1}^\infty \left(a^\alpha_{-m}a^\alpha_m + \frac{1}{2}[a_m^\alpha, a^\alpha_{-m}]\right)\right)$$

$$= L_0^\perp + \frac{1}{2}(D-2)\sum_{m=1}^\infty m = L_0^\perp + \frac{D-2}{2}\zeta(-1) = L_0^\perp - \frac{D-2}{24}. \tag{5.24}$$

In the last steps we have regularized the divergent infinite sum by considering it as the analytic continuation of the RIEMANN ζ-function,

$$\sum_{m=1}^{\infty} m^{-s} = \zeta(s), \tag{5.25}$$

which is well-defined for $Res > 1$. The analytic continuation to $s = -1$ is unambiguous and gives the value $\zeta(-1) = -\frac{1}{12}$. Thereby, we conclude that the normal ordering constant a is related to the dimension of the ambient space by

$$a = \frac{D-2}{24}. \tag{5.26}$$

Let us note that the state space $\mathcal{H}_{\mathscr{A}}$ we have defined above is manifestly positive definite. Recall that the negative norm states in the covariant construction came with the creation operators a_{-n}^0 for oscillations in the time direction $\mu = 0$. In light-cone gauge, we have only transverse oscillators. Since the metric in the $(D-2)$-dimensional transverse space spanned by x^{α} is Euclidean, $\mathcal{H}_{\mathscr{A}}$ is a HILBERT space. The statement is actually true regardless of our choice of the constant a and the dimension D. One may therefore wonder whether there is any restriction on the dimension D of the light-cone gauged string theory.

5.4 LORENTZ Symmetry and 26 Dimensions

Of course, the restriction to $D = 26$ is not just an artifact of the covariant quantization. It also exists in the light-cone quantization, but the mechanism by which it arises is quite different. While discussing the particle theory we have put a lot of stress on the issue of symmetries only to find that the state space of the relativistic particle carries an action of the POINCARÉ algebra even though the light-cone gauge condition did not preserve all space-time symmetries. As we shall argue now, the stringy analogue of this discussion is much more interesting. It turns out that the state space $\mathcal{H}_{\mathscr{A}}$ constructed in the previous section carries an action of the POINCARÉ algebra if and only if $D = 26$ (and $a = 1$).

In order to arrive at this conclusion, we consider once more the generators $M^{\mu\nu}$ of boosts and rotations. In our bosonic field theory, the NOETHER charges for Lorentz transformations can expressed as

$$M^{\mu\nu} = \int_0^{2\pi} d\sigma \, (X^{\mu}(\sigma,0)\Pi^{\nu}(\sigma,0) - \Pi^{\mu}(\sigma,0)X^{\nu}(\sigma,0))$$

$$= x^{\mu}p^{\nu} - p^{\mu}x^{\nu} - i\sum_{n=1}^{\infty}\frac{1}{n}(a_{-n}^{\mu}a_n^{\nu} - a_{-n}^{\nu}a_n^{\mu}) - i\sum_{n=1}^{\infty}\frac{1}{n}(\bar{a}_{-n}^{\mu}\bar{a}_n^{\nu} - \bar{a}_{-n}^{\nu}\bar{a}_n^{\mu}). \tag{5.27}$$

Here, Π^μ denotes the momentum $\Pi^\mu(\sigma, \tau) = (2\pi\alpha')^{-1}\partial_0 X^\mu(\sigma, \tau)$ conjugate to the coordinate field X^μ. In equation (5.27), the generator $M^{-\alpha}$ is expressed through operators of the unreduced theory. According to our general prescription, we are instructed to set $x^+ = 0 = a_n^+ = \bar{a}_n^+$ and to replace $a_n^-, \bar{a}_n^-$, and p^- through the expressions in eqs. (5.16)–(5.18). For the generators $M^{-\alpha}$ the result is given by

$$M^{-\alpha} = x^- p^\alpha - \frac{1}{\alpha' p^+} x^\alpha l_0^\perp - \frac{i}{\sqrt{2\alpha'}} \frac{1}{p^+} \sum_{n \neq 0} \frac{1}{n} l_{-n}^\perp a_n^\alpha - \frac{i}{\sqrt{2\alpha'}} \frac{1}{p^+} \sum_{n \neq 0} \frac{1}{n} \bar{l}_{-n}^\perp \bar{a}_n^\alpha. \quad (5.28)$$

Now we have to quantize our formula for $M^{-\alpha}$. This turns all coordinates into the usual operators. The functions $l_n^\perp$, in particular, are replaced by $L_n^\perp - a\delta_{n,0}$. Furthermore, the symmetric ordering prescription should be applied to resolve ordering ambiguities. After these steps we arrive at

$$M^{-\alpha} = x^- p^\alpha - \frac{1}{\alpha' p^+} \left(x^\alpha (L_0^\perp - a) + (L_0^\perp - a) x^\alpha \right) - \frac{i}{\sqrt{2\alpha'}} \frac{1}{p^+} \sum_{n=1}^{\infty} \frac{1}{n} (L_{-n}^\perp a_n^\alpha - a_{-n}^\alpha L_n^\perp). \quad (5.29)$$

Here, we have omitted all terms involving $\bar{a}_n^\alpha$. As we have pointed out in our discussion of the particle theory, the state space can carry only a representation of the Lorentz group $SO(1, D-1)$ if the commutator of the generators $M^{-\alpha}$ vanishes, $[M^{-\alpha}, M^{-\beta}] = 0$. A straightforward but cumbersome calculation leads to the following answer (see, e.g., [2] for a full derivation):

$$[M^{-\alpha}, M^{-\beta}] = -\frac{1}{\alpha'(p^+)^2} \sum_{m=1}^{\infty} \left(a_{-m}^\alpha a_m^\beta - a_{-m}^\beta a_m^\alpha \right) \left\{ m \left[1 - \frac{D-2}{24} \right] + \frac{1}{m} \left[\frac{D-2}{24} - a \right] \right\}. \quad (5.30)$$

Once more we did not display the terms that contain $\bar{a}_n^\alpha$. Hence, in order for the $M^{\mu\nu}$ to satisfy the relations of a Lorentz algebra, we need to impose the conditions

$$D = 26 \quad \text{and} \quad a = 1. \quad (5.31)$$

Other generators $M^{\mu\nu}$ can be constructed following the same steps we described for $M^{-\alpha}$. Their commutators turn out to obey the relations of the Lorentz-algebra for the same values of D and a. Since $M^{\nu\mu}$ also commute with the quantum constraint $L_0 - \bar{L}_0$, we conclude that our state space $\mathcal{H}_{\mathscr{A}}$ carries an action of $SO(1, D-1)$ if and only if $D = 26$ and $a = 1$. For this distinguished choice of parameters, the state space $\mathcal{H}_{\mathscr{A}}$ can be shown to agree with the state space we constructed through the covariant quantization of the closed bosonic string. Note that the tachyonic and massless states are obviously the same in both constructions. Checking the equivalence of the two quantization procedures for the entire spectrum of vibrational modes is much more involved and beyond the scope of this chapter; see, e.g., [27] for a more detailed discussion and references to the original literature.

Exercises

Problem 12. *Using light-cone quantization for closed bosonic strings, describe a complete basis of eigenstates of the mass operator M^2 with eigenvalue $M^2 = 4/\alpha'$.*

Problem 13. *In light-cone quantization the generators of* LORENTZ *transformations $M^{-\alpha}$ are given by eq. (5.29). Compute the commutator $[M^{-\alpha}, M^{-\beta}]$.*

6

Branes and Quantization of Open Strings

We shall now discuss how to model open bosonic strings. Most of the steps follow very closely the program we carried out for closed strings. Therefore, our discussion will be rather brief, focusing on the main new elements of the theory. The principal goal of the chapter is to construct the wave functions of open bosonic strings, and in particular its massless states. The latter are shown to include a massless gauge boson.

6.1 Boundary Conditions for Bosonic Strings

For the open string, the parametrization fields $X^\mu(\tau, \sigma)$ map the strip $\mathbb{R} \times [0, \pi]$ into the space-time manifold $\mathcal{M}$; see Figure 6.1. The motion of open bosonic strings admits several equivalent classical descriptions through an open version of the NAMBU-GOTO action, a POLYAKOV type action functional, or as a free field theory with constraints. Here we shall adopt the third approach right away. Formally, the action looks the same as before:

$$S[X] = -\frac{1}{4\pi\alpha'} \int d\tau d\sigma \, \partial_a X^\mu \partial^a X_\mu. \tag{6.1}$$

In comparison to the case of closed strings, we now integrate over the infinite strip $\Sigma = \mathbb{R} \times [0, \pi]$. The constraints are identical to those we derived for closed strings:

$$\partial_a X^\mu \partial_b X_\mu = \frac{1}{2}\eta_{ab}\partial_c X^\mu \partial^c X_\mu. \tag{6.2}$$

A moment of reflection shows that our classical problem is not properly posed yet due to the presence of the boundary $\partial\Sigma$. We want solutions of the 2-dimensional wave equation on the strip Σ to extremize the action S. The explicit variation of

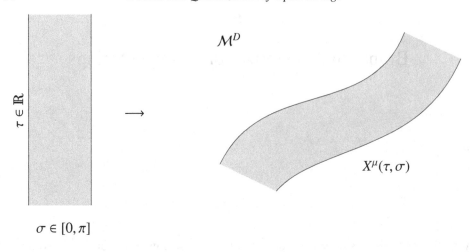

Figure 6.1 The world-sheet of an open string is a map from the strip $\Sigma = \mathbb{R} \times [0, \pi]$ into the D-dimensional space-time manifold.

$S[X]$ shows that this is the case only if we impose appropriate boundary conditions at $\sigma = 0, \pi$. In fact, the variation and consecutive integration by parts give

$$\delta S[X] = \frac{1}{2\pi\alpha'} \int_{\mathbb{R}} d\tau \int_0^\pi d\sigma \left(\partial_\tau X^\mu \partial_\tau \delta X_\mu - \partial_\sigma X^\mu \partial_\sigma \delta X_\mu \right)$$

$$= -\frac{1}{2\pi\alpha'} \int d\tau d\sigma \left(\partial_\tau^2 X^\mu - \partial_\sigma^2 X^\mu \right) \delta X_\mu - \frac{1}{2\pi\alpha'} \int_{\mathbb{R}} d\tau \, \partial_\sigma X^\mu \delta X_\mu \bigg|_0^\pi .$$

$$(6.3)$$

While the bulk term vanishes on all solutions of the wave equation, as requested, the unwanted boundary contributions can be avoided only by imposing either of the following two boundary conditions. One possibility is to fix the value of the field X^μ at the boundary. Alternatively, we can also require the σ-derivative of X^μ to vanish. We shall assume that the latter is realized for the time component X^0. Then, after a possible re-labeling of directions, we have

- NEUMANN boundary conditions: $\partial_\sigma X^\mu(\tau, \sigma)|_{\sigma \in \{0,\pi\}} = 0$ for $\mu \in \{0, 1, \ldots, p\}$,
- DIRICHLET boundary conditions: $X^\mu(\tau, \sigma)|_{\sigma \in \{0,\pi\}} = x_0^\mu$ for $\mu \in \{p+1, \ldots, D-1\}$.

The geometrical interpretation of these boundary conditions is straightforward. For the first $p + 1$ coordinates, the string ends are moving freely. On the other hand, both ends are fixed in the remaining $D - p - 1$ directions transverse to the $(p + 1)$-dimensional hyperplane defined by $x^\mu = x_0^\mu$ for $\mu \in \{p + 1, \ldots, D - 1\}$. These boundary conditions break the $SO(1, D - 1)$ symmetry group of the bulk down to the LORENTZ group $SO(1, p)$ in $p + 1$ dimensions. Originally, such a reduction of the space-time symmetry was considered unphysical, and consequently NEUMANN boundary conditions were always imposed in all D directions of bosonic string theory. But the discovery of branes has changed the perspective on such issues. Recall from the first chapter that branes are higher dimensional solitons of string theory that extend along a $(p + 1)$-dimensional hyperplane. Placing such objects into MINKOWSKI space breaks $SO(1, D - 1)$ down to $SO(1, p) \times SO(D - p - 1)$ just as our boundary conditions. We shall see later that boundary conditions of the above form indeed model a large class of branes in superstring theory. Before we conclude, let us remark that our boundary conditions are not the most general ones. In our geometric terms we have assumed that the open string has both its ends on the same brane. But there can be more than one brane present, in which case the string's two ends may be located on different branes. This suggests allowing e.g. the components X^μ, $\mu = p - 1, \ldots, D$ to assume different values on the two boundaries in case the branes' positions along the μth coordinate direction is different. It may also be necessary to impose NEUMANN boundary conditions on X^μ at one of the boundary points and DIRICHLET boundary conditions at the other. Thereby we capture a situation where one of the two branes that our open string connects extends along the μth direction while the other does not. Although certainly of great interest, we shall not consider this second possibility throughout these chapters. On the other hand, we shall allow for different values of the field X^μ, $\mu > p$, at the two endpoints of the string.

6.2 Quantization with NEUMANN Boundary Conditions

We would now like to solve the initial value problem for open strings, i.e., find solutions for the 2-dimensional wave equations with the appropriate boundary conditions. As usual, a general solution of the wave equation may be written as a sum of a left- and a right-moving term. While left and right movers were independent for closed strings, they are reflected into each other on the boundary of an open string world-sheet. We have decided to impose NEUMANN boundary conditions in the directions of $\mu \in \{0, 1, \ldots, p\}$ so that the corresponding solution reads

$$X^\mu(z, \bar{z}) = x^\mu - i\alpha' p^\mu \ln(z\bar{z}) + i\sqrt{\frac{\alpha'}{2}} \sum_{n \neq 0} \frac{a_n^\mu}{n} \left(z^{-n} + \bar{z}^{-n}\right). \tag{6.4}$$

Note that there is only one set of oscillators a_n^μ appearing in eq. (6.4), in line with our expectations. In agreement with earlier conventions, we are working with the world-sheet variables z and $\bar{z}$ instead of the EUCLIDEAN time τ and σ. Both sets of coordinates are related by a simple conformal map (3.9) from the strip $\mathbb{R} \times [0, \pi]$ to the upper half of the complex plane (see Figure 6.2),

$$z = e^{\tau - i\sigma} \quad \text{and} \quad \bar{z} = e^{\tau + i\sigma}. \tag{6.5}$$

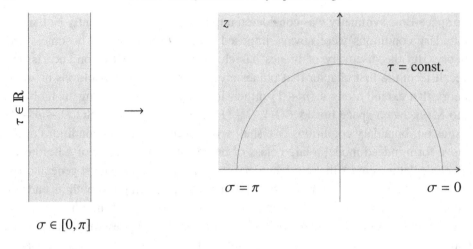

$\sigma \in [0, \pi]$

Figure 6.2 The conformal transformation $z = e^{\tau - i\sigma}$ maps the strip Σ to the upper half of the complex plane.

In order to check that the solution X^μ in eq. (6.4) does indeed satisfy NEUMANN boundary conditions, we calculate the σ-derivative of X^μ along the line $z = \bar{z}$:

$$\partial_\sigma X^\mu|_{z=\bar{z}} = i\sqrt{\frac{\alpha'}{2}} \sum_{n\neq 0} a_n \left(i\,z^{-n} - i\,\bar{z}^{-n} \right) \Bigg|_{z=\bar{z}} = 0. \qquad (6.6)$$

Following the same arguments we outlined for closed strings, it is easy to see that the coordinates a_n^μ, x^μ, and p^μ obey the familiar POISSON algebra. After quantization we obtain conjugate operators x^μ, p^μ, and a single set of oscillators a_n^μ. For convenience we define $a_0^\mu := \sqrt{2\alpha'}p^\mu$. Note that the normalization differs from the closed string case by an extra factor of 2.

Once more, the state space of the free field theory is generated from a family of ground states $|k\rangle$, $k = (k^0, k^1, \ldots, k^p)$, with the defining properties

$$a_n^\mu |k\rangle = 0 \quad \text{for all } n > 0 \ , \quad p^\mu |k\rangle = k^\mu |k\rangle \quad \text{for } \mu \in \{0, 1, \ldots, p\}$$

by application of creation operators $a_n^\mu, n < 0$. In this way we construct the elementary building blocks

$$\mathcal{H}_k = \text{span}\left\{ a_{n_1}^{\mu_1} \cdots a_{n_s}^{\mu_s} |k\rangle \ | \ n_1 \leq n_2 \leq \cdots \leq n_s < 0; \mu_1, \mu_2, \ldots, \mu_s \in \{0, 1, \ldots, p\} \right\}. \qquad (6.7)$$

From these, the total state space is again put together as a direct integral:

$$\mathcal{H}_{Dp,N}^{OBF} = \oint d^{p+1}k\, \mathcal{H}_k. \qquad (6.8)$$

It has the same form as for the free bosonic field theory on a cylinder, except that there is only one set of oscillators a_n^μ rather than two for the independent left and right movers of closed strings.

With the state space being constructed, we can start to build the various fields that are acting on it. Here, we shall be most interested in the fields that we can insert at points $u \in \mathbb{R}$ on the boundary of our world-sheet. Such fields turn out to be in one-to-one correspondence with states in the space (6.8), very much in the same way as for the theory on the cylinder before. The only fields we shall need later are the vertex operators that come with states of the form $a_{-1}^\mu \, | k \rangle$. These are built as a normal ordered product from the exponential fields $V_k(u), u = Re\, z$, and derivatives of $X^\mu(u)$:

$$a_{-1}^\mu \, | k \rangle = \, : i\, \partial_u X^\mu(0)\, V_k(0) : \, | 0 \rangle , \qquad (6.9a)$$

$$\text{where} \quad V_k(u) := \, : \exp\left(i k \cdot X(u)\right) : . \qquad (6.9b)$$

These vertex operators $V_k(u)$ themselves are associated to the ground states $| k \rangle$ in the sense that $V_k(0) \, | 0 \rangle = | k \rangle$.

6.3 Quantization with DIRICHLET Boundary Conditions

Having quantized the first $p + 1$ components of our free bosonic field with NEU-MANN boundary conditions, we now need to consider the remaining directions for which we have decided to impose DIRICHLET boundary conditions. We shall extend the previous setting slightly by allowing the endpoints at $\sigma = 0$ and $\sigma = \pi$ to assume different values,

$$X^\alpha(\tau, 0) = x_0^\alpha \qquad \text{and} \qquad X^\alpha(\tau, \pi) = x_1^\alpha, \qquad (6.10)$$

where $\alpha = p + 1, \ldots, D - 1$ labels the directions transverse to the two hyperplanes defined by $x^\alpha = x_{0,1}^\alpha$. Once more, the general solution to the wave equation may be written as a sum of left- and right-moving terms. DIRICHLET boundary conditions determine left movers, in terms of right movers and they constrain the zero modes,

$$X^\alpha(z, \bar{z}) = x_0^\alpha - i\alpha' \Delta^\alpha \ln\left(\frac{z}{\bar{z}}\right) + i\sqrt{\frac{\alpha'}{2}} \sum_{n \neq 0} \frac{a_n^\mu}{n} \left(z^{-n} - \bar{z}^{-n}\right), \qquad (6.11)$$

where x_0^α and the combination

$$\Delta^\alpha = \frac{x_0^\alpha - x_1^\alpha}{2\pi \alpha'}$$

that appears instead of the momentum operator p^μ are completely determined by the boundary values of the fields X^α. It is easy to see that the solution (6.11) satisfies the boundary conditions (6.10). As before, the coordinates a_n^α satisfy the standard POISSON relations, and the POISSON commute with the numbers x_0^α and Δ^α. After quantization, the oscillators a_n^α with $n \neq 0$ become operators, while $x^\alpha = x_0^\alpha$ and $a_0^\alpha = \sqrt{2\alpha'}\Delta^\alpha$ remain numbers.

The state space of our field theory is generated from a single ground state $|\Delta\rangle$, by application of the creation operators $a_n^\alpha, n < 0$:

$$\mathcal{H}_{Dp,D}^{OBF} = \mathcal{H}_\Delta$$
$$= \left\{ a_{n_1}^{\alpha_1} \cdots a_{n_s}^{\alpha_s} |\Delta\rangle : n_1 \leq n_2 \leq \cdots \leq n_s < 0; \alpha_i \in \{p+1, \ldots, D-1\} \right\}.$$
$$(6.12)$$

For completeness let us also spell out that the ground state $|\Delta\rangle$ possesses the defining properties

$$a_n^\alpha |\Delta\rangle = 0 \quad \text{for all } n > 0, \tag{6.13}$$

$$a_0^\alpha |\Delta\rangle = \sqrt{2\alpha'}\Delta^\alpha |\Delta\rangle, \tag{6.14}$$

where $\alpha \in \{p+1, \ldots, D-1\}$. On the state space (6.12) we can act with the usual derivative fields $J^\mu(u), u \in \mathbb{R}$.

6.4 Mass Spectrum of the Open Bosonic String

The total state space of the bosonic field theory on the strip is given as a direct product of the two sectors with NEUMANN and DIRICHLET boundary conditions, thereby describing the directions both along and perpendicular to our Dp-branes:

$$\mathcal{H}_{Dp}^{OBF} = \mathcal{H}_{Dp,N}^{OBF} \otimes \mathcal{H}_{Dp,D}^{OBF}. \tag{6.15}$$

Let us stress once more that the indices μ on the oscillators a_n^μ run over all directions $\mu = 0, \ldots, D-1$ of the target space, whereas the vector k is a $p+1$ vector only. As in the case of closed strings, the vector k is going to parametrize the center of mass momentum of a string. But since the ends of our strings are constrained by $X^\alpha(\tau, 0) = x_0^\alpha$ and $X^\alpha(\tau, \pi) = x_1^\alpha, \alpha = p+1, \ldots, D-p-1$, the center of mass momentum must be tangential to this hypersurface. In between the end-points, however, the open string is allowed to leave the world-volume of the Dp-brane and hence can fluctuate in all target space dimensions.

One of the fields that act on the state space (6.15) of the free field theory is the stress energy tensor T_{ab}. On the cylinder, the latter had two independent components T and $\bar{T}$ associated with left and right movers. In the present context, left and right movers are related by a reflection on the boundary, and hence the modes of the stress energy tensor give rise to the action of only one VIRASORO algebra on the state space. It may be constructed from the one set of oscillators a_n^μ through the familiar formula

$$L_n = \frac{1}{2} \sum_{m \in \mathbb{Z}} : a_m^\mu a_{n-m}^\nu : \eta_{\mu\nu} + \frac{1}{2} \sum_{m \in \mathbb{Z}} : a_m^\alpha a_{n-m}^\beta : \delta_{\alpha\beta}, \qquad (6.16)$$

where μ, ν run from 0 to p and α, β run from $p+1$ to $D-1$. Recall that the elements a_0^α are simply numbers. The L_n satisfy the defining relations of the VIRASORO algebra with the same central charge $c = D$ as on the cylinder. For later use we want to spell out the generator L_0 a bit more explicitly:

$$L_0 = -\alpha' M^2 + \alpha' \Delta^2 + \sum_{m>0} : a_{-m}^\mu a_m^\nu : \eta_{\mu\nu} + \sum_{m>0} : a_{-m}^\alpha a_m^\beta : \delta_{\alpha\beta}. \qquad (6.17)$$

Here, the operator M^2 stands for $M^2 = -p^2 = \sum_{\mu=0}^p p^\mu p_\mu$, and Δ is the transverse distance between the two branes.

Now we have all the ingredients that are necessary to define and analyze the state space for the open bosonic string. As before, we need to impose our constraints and then remove all null states:

$$\mathcal{H}_{\mathrm{D}p}^{\mathrm{OBS}} = \left\{ |\psi\rangle \in \mathcal{H}_{\mathrm{D}p}^{\mathrm{OBF}} : L_n |\psi\rangle = 0 \text{ for } n > 0; L_0 |\psi\rangle = |\psi\rangle \right\} / \mathcal{N}. \quad (6.18)$$

According to the result of BROWER-GODDARD-THORN, the state space (6.18) is positive definite for $D = 26$ regardless of the dimensionality of our brane. Note that the number $p+1$ of directions with NEUMANN boundary conditions has no influence beyond the zero modes of the theory. Since all potential negative norm states are associated with excited states, their existence is independent of p.

With the space of wave function being well defined for $D = 26$, we would like to analyze the mass spectrum of the open bosonic string. Note that the mass operator M^2 commutes with all constraints so that it descends to wave functions of the string. Explicitly, it is given by

$$M^2 = -\frac{1}{\alpha'} + \Delta^2 + \frac{1}{\alpha'} \sum_{n=1}^\infty n N_n. \qquad (6.19)$$

Once more, the lowest eigenvalue of M^2 is negative, at least for Δ^2 small enough. In other words, the spectrum of open bosonic strings stretched between our two branes contains a scalar tachyon whose wave function is a sum of states

$$|\phi; k\rangle_\Delta = \hat{\phi}(k) |k\rangle_\Delta = \hat{\phi}(k) |k\rangle \otimes |\Delta\rangle, \qquad (6.20)$$

with some $p+1$ momentum $k = (k_0, \ldots, k_p)$. The factor $\hat{\phi}(k)$ is the amplitude of the momentum eigenstates, and we are looking for the constraints our physical state conditions impose on $\hat{\phi}(k)$. To begin, our state $|\phi\rangle$ is restricted by the mass shell condition

$$L_0 |\phi;k\rangle_\Delta = |\phi;k\rangle_\Delta \quad \Leftrightarrow \quad k^2 \hat{\phi}(k) = \left(\frac{1}{\alpha'} - \Delta^2\right) \hat{\phi}(k). \tag{6.21}$$

This means that $\hat{\phi}(k)$ must vanish away from the hyperboloid $k^2 = 1/\alpha' - \Delta^2$. If we pass from momentum to coordinate space, the same condition reads

$$\left(\Box + \frac{1}{\alpha'} - \Delta^2\right) \phi(x) = 0, \tag{6.22}$$

i.e., the lowest mass particle in our spectrum satisfies the KLEIN-GORDON equation with mass $M^2 = \Delta^2 - 1/\alpha'$. While the mass is negative for vanishing transverse distance between the two branes, we can make it positive by increasing the distance to some value larger than the string length. This outcome is rather natural: the dependence of the mass on the separation between the branes originates from the string's tension. Stretching it some distance Δ costs energy. The latter is perceived as a mass shift of the particle spectrum.

Let us now turn our attention to the first excited level. Once more we begin with the most general state from the total state space of the field theory. Its component with $p + 1$ momentum k has the form

$$|\xi,k\rangle_\Delta = \left(\xi_\mu(k) a^\mu_{-1} + \xi_\alpha(k) a^\alpha_{-1}\right) |k\rangle \otimes |\Delta\rangle. \tag{6.23}$$

Here, $\xi^\mu = \xi^\mu(k)$ and $\xi^\alpha = \xi^\alpha(k)$ are the polarizations along and transverse to the brane. By imposing the physical state conditions we constrain the polarizations as follows:

$$(L_0 - 1) |\xi,k\rangle_\Delta = 0 \quad \Leftrightarrow \quad k_\mu k^\mu \xi^{\nu,\alpha}(k) = -\Delta^2 \xi^{\nu,\alpha}(k), \tag{6.24a}$$

$$L_1 |\xi,k\rangle_\Delta = 0 \quad \Leftrightarrow \quad \xi_\mu(k) k^\mu + \xi_\alpha(k) \Delta^\alpha = 0. \tag{6.24b}$$

Additionally, we have to check whether the state is null:

$$_\Delta \langle \xi,k | \xi,k\rangle_\Delta = 0 \quad \Leftrightarrow \quad \xi_\mu(k) \xi^\mu(k) + \xi_\alpha(k) \xi^\alpha(k) = 0. \tag{6.24c}$$

Let us rewrite the outcome of our short analysis for $\Delta = 0$ in terms of coordinate space fields. We shall denote the FOURIER transform of $\xi^\mu(x)$ by $A^\mu(x)$ for $\mu = 0,\ldots,p$ and introduce the symbol Φ^α to denote the FOURIER transform of the components $\xi^\alpha(k)$. Note that x is a coordinate on the $(p + 1)$-dimensional world-volume of the brane, not in the 26-dimensional background. The first equation in the above list tells us that both A^μ and Φ^α satisfy the $(p + 1)$-dimensional wave equation:

$$\Box_p A_\mu(x) = 0 \quad , \quad \Box_p \Phi^\alpha = 0. \tag{6.25}$$

Continuing to assume that $\Delta = 0$, the second condition states that ξ^μ is transverse or in coordinate space:

$$\partial_\mu A^\mu = 0. \tag{6.26}$$

There is no condition on the polarization ξ^α and hence on the fields Φ^α. Finally, the only null states come with the directions parallel to the branes. They appear when the polarization ξ^μ is parallel to the momentum k^μ or in terms of coordinate space fields if the vector field is "pure gauge," i.e., $A_\mu = \partial_\mu \Lambda$. We conclude that the field A^μ is a massless vector particle with $p - 1$ degrees of freedom. The equations we found for A^μ coincide with those for a photon or MAXWELL field in LORENTZ gauge. In addition, $D - p - 1$ massless scalar fields Φ^α are propagating on the world-volume of a Dp-brane, one for each transverse direction. We note that the massless scalar fields Φ^α are in one-to-one correspondence with the parameters x_0^α that specify the brane's transverse position. As we proceed we shall see many more examples of this relation between massless scalars and geometric parameters (moduli) of the string's background.

Exercises[1]

Problem 14. *(a) Determine the equations of motion from the following classical action:*

$$S[X] = -\frac{1}{4\pi\alpha'} \int d\tau d\sigma \left(\eta^{ab} \partial_a X^\mu \partial_b X^\nu \eta_{\mu\nu} + \epsilon^{ab} \partial_a X^\mu \partial_b X^\nu B_{\mu\nu} \right).$$

on the cylinder with periodic boundary conditions. Here, $B_{\mu\nu}$ denote the elements of a constant anti-symmetric matrix B.

(b) Consider the action from part (a) on a the strip $\sigma \in [0, \pi], \tau \in R$. Show that the boundary terms in the variation of the action vanish if one imposes the following boundary condition:

$$\partial_\sigma X^\mu = \eta^{\mu\nu} B_{\nu\rho} \partial_\tau X^\rho \big|_{\sigma=0,\pi}. \tag{6.27}$$

Problem 15. *(a) Show that the general solution of the 2-dimensional wave equation with boundary conditions (6.27) can be written in the form*

$$X^\mu(z, \bar{z}) = \hat{x}^\mu - i\sqrt{\frac{\alpha'}{2}} \alpha_0^\mu \ln z\bar{z} - i\sqrt{\frac{\alpha'}{2}} B^\mu{}_\nu \alpha_0^\nu \ln \frac{z}{\bar{z}}$$

$$+ i\sqrt{\frac{\alpha'}{2}} \sum_{n \neq 0} \frac{\alpha_n^\mu}{n} \left(z^{-n} + \bar{z}^{-n} \right) + i\sqrt{\frac{\alpha'}{2}} \sum_{n \neq 0} \frac{B^\mu{}_\nu \alpha_n^\nu}{n} \left(z^{-n} - \bar{z}^{-n} \right),$$

where summation over $\nu = 1, \ldots, p$, is understood.

---

[1] The first few exercises are connected and point at a very interesting connection between open strings and non-commutative geometry; see [15, 59, 60].

(b) Show that the set of commutation relations

$$[\alpha_n^\mu, \alpha_m^\nu] = n\, G^{\mu\nu}\, \delta_{n,-m}, \qquad [\hat{x}^\mu, \alpha_n^\nu] = i\sqrt{\alpha'}\, G^{\mu\nu}\, \delta_{0,n},$$

$$[\hat{x}^\mu, \hat{x}^\nu] = i\,\Theta^{\mu\nu}$$

with

$$G^{\mu\nu} = \left(\frac{1}{g+B}\right)_S^{\mu\nu}, \qquad \Theta^{\mu\nu} = \left(\frac{\alpha'}{g+B}\right)_A^{\mu\nu}$$

guarantee that the fields X and $\dot{X}$ satisfy the canonical commutation relations

$$[\dot{X}^\mu(\tau,\sigma), X^\nu(\tau,\sigma')] = 2\pi i \alpha' \eta^{\mu\nu} \delta(\sigma - \sigma').$$

Problem 16. *(a) Let $\hat{x}^\nu, \nu = 1, \ldots, p$, be a set of operators satisfying $[\hat{x}^\mu, \hat{x}^\nu] = i\Theta^{\mu\nu}$. Compute the product of exponentials*

$$e^{ik_\mu \hat{x}^\mu} e^{ik'_\nu \hat{x}^\nu}.$$

(b) For a given function f define an operator $f(\hat{x})$ by the prescription

$$f(\hat{x}) := \int d^d k \hat{f}(k) \exp(ik_\nu \hat{x}^\nu),$$

where $\hat{f}$ denotes the Fourier transform of f. Given any two functions f and g, determine their MOYAL-WEYL *product $f * g(\hat{x}) := f(\hat{x})g(\hat{x})$.*

Problem 17. *Calculate the correlation function of vertex operators*

$$\psi_k(u) := \; : e^{ik_\mu X^\mu(u)} : \; = \; e^{ik_\mu X_<^\mu(u)}\, e^{ik_\mu X_>^\mu(u)},$$

where

$$X_>^\mu(u) = -i\sqrt{2\alpha'}\, \alpha_0^\mu \ln u + i\sqrt{2\alpha'} \sum_{n>0} \frac{\alpha_n^\mu}{n} u^{-n},$$

$$X_<^\mu(u) = \hat{x}^\mu + i\sqrt{2\alpha'} \sum_{n<0} \frac{\alpha_n^\mu}{n} u^{-n}.$$

and u assumes values along the real line $u \in \mathbb{R}$ by following the steps explained in Section 3.3.

Problem 18. *Construct the most general solution of the 2-dimensional wave equation on a strip with mixed boundary conditions*

$$\partial_\tau X(\tau,\sigma)|_{\sigma=0} = 0, \qquad \partial_\sigma X(\tau,\sigma)|_{\sigma=\pi} = 0.$$

7

Open Strings and Gauge Theory

In the last chapter we saw that among the massless open string modes on a Dp-brane one can find a vector particle with $p-1$ transverse degrees of freedom. We claimed in the introduction that these scatter like gauge bosons, at least in the limit when the string length becomes small. Now we have assembled enough background to check this claim. In this chapter we will state the rule for computing tree-level scattering amplitudes for open strings. We shall first do that for a single brane and then generalize to stacks of N identical branes. After a short interlude on non-Abelian gauge theory we shall compute the 3-vertex of open strings and compare with the corresponding quantity for gauge bosons.

7.1 Tree-Level Scattering Amplitudes for Open Strings

A string scattering amplitude assigns a number to a set of states $\psi_i \in \mathcal{H}_{\text{D}p}^{\text{OBS}}$ of the open bosonic string in a Dp-brane background. The prescription resembles the one we have used for closed strings, with some additional complications. Before we get there, let us recall that any open string scattering amplitude with M external legs and without holes can be cut into $M-2$ 3-vertices. The latter are weighted with a coupling g_o, and hence the M-point vertex is proportional g_o^{M-2}. Once more, we want all tree-level amplitudes to come with the same power of g_o. To this end, we absorb one factor g_o into each external leg. After this change of normalization, an open string tree-level amplitude receives a factor $g_o^{-2} = g_s^{-1}$. The unique string tree-level FEYNMAN graph with M external legs is depicted in Figure 7.1.

As in the case of closed strings, the computation of string scattering amplitudes is reduced to the computation of correlation functions in the corresponding worldsheet (free) field theory. To this end, we make use of the state field correspondence, which implies that for any open string wave function ψ_i there exists a unique field $V_{\psi_M} = V_M$ such that $V_M(0)|0\rangle = |\psi_M\rangle$. For wave functions of open strings, these fields are inserted somewhere along the boundary of the upper half-plane, i.e., at

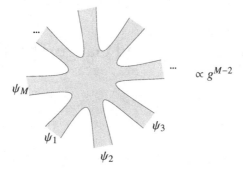

Figure 7.1 An open tree-level diagram with M external legs.

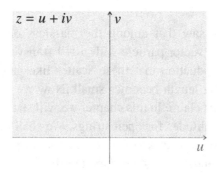

Figure 7.2 $u = Re(z)$ can be used to label the boundary of the open string world-sheet.

points $z = u \in \mathbb{R}$ (see Figure 7.2). Since the insertion point along the boundary does not correspond to any target space data, it needs to be integrated over

$$A(\psi_1, \dots, \psi_M) \sim g_s^{-1} \langle \mathcal{O}_{\psi_1} \cdots \mathcal{O}_{\psi_M} \rangle_{\mathrm{FT}}, \qquad (7.1)$$

where $\mathcal{O}_\psi$ are integrated vertex operators:

$$\mathcal{O}_\psi = \int_{\mathbb{R}} du \, V_\psi(u). \qquad (7.2)$$

If we are interested in the amplitude of open string tachyons, for example, the external states are given by $|k_i\rangle$ with $k_i^2 = 1/\alpha$, and the corresponding fields are our usual exponentials:

$$V_\psi(u) = : \exp(i k \cdot X(u)) : \quad \text{where} \quad k^2 = 1/\alpha', \qquad (7.3a)$$

$$X^\mu(u) = x^\mu - 2i \alpha' p^\mu \ln(u) + i \sqrt{2\alpha'} \sum_{n \neq 0} \frac{a_n^\mu}{n} u^{-n}. \qquad (7.3b)$$

The definition (7.1) is morally correct but suffers from two little defects that still need to be cured before we have a well-defined construction of scattering amplitudes.

The first problem originates from the symmetries of the correlation functions. A close inspection shows that they are invariant under all rational transformations of the real line, i.e., under all the maps

$$u \mapsto \frac{au + b}{cu + d} \quad \text{where} \quad ad - bc = 1.$$

Here all parameters a, b, c, d are taken to be real, and the rational transformations we described provide an action of the non-compact group SL(2,$\mathbb{R}$) on the real line. Note that such transformations include the simple translations $u \mapsto u + b$. Because of these symmetries, the amplitude $\mathcal{A}$ we defined above is proportional to the volume of SL(2,$\mathbb{R}$). In order to get something finite we need to divide by this volume. Since the three-parameter group of rational transformations can be used to put three insertion points into $u_1 = \infty, u_2 = 0, u_3 = 1$, the correct prescription is again to fix these three insertions and then to integrate only over the remaining $M - 3$ insertion points $u_4, \ldots, u_M$. This will be part of the final formula for $\mathcal{A}$ below.

Before we get there, let us address another problem with our prescription (7.1). It originates from the fact that correlation functions on the boundary of the half-plane such as

$$\langle V_{k_1}(u_1) \cdots V_{k_M}(u_M) \rangle = \delta \left(\sum_{i=1}^{M} k_i \right) \prod_{i<j} (u_i - u_j)^{2\alpha' k_i \cdot k_j} \tag{7.4}$$

are well defined only for $u_1 > u_2 > \cdots > u_M$. In the case of closed string amplitudes we could safely drop such a condition because the correlator depends only on $|z_i - z_j|$ and hence possesses a unique analytic continuation from $|z_i| > |z_{i+1}|$ to the full complex plane. The correlation functions (7.4) do contain arbitrary powers $(u_i - u_j)^\kappa$ where κ does not need to be an even integer. Hence, there is no way to extend such correlators to N unordered points on the real line. In the computation of string amplitudes we want to integrate all insertion points $u_4, \ldots, u_M$ over the entire real line. What we have to keep in mind is that the field theory correlators need to be re-evaluated whenever the order of the insertion points changes.

After this preparation we are now ready to give a precise formula for how we want to compute tree-level amplitudes for open strings on a single Dp-brane. It is given by

$$\mathcal{A}(\psi_1, \ldots, \psi_M) = g_s^{-1} \int du_4 \cdots du_M \sum_{\sigma \in S^{M-1}} \Theta \left(u_{\sigma(2)}, \ldots, u_{\sigma(M)} \right)$$
$$\langle V_1(\infty) V_{\sigma(2)}(u_{\sigma(2)}) \cdots V_{\sigma(M)}(u_{\sigma(M)}) \rangle, \tag{7.5}$$

where $u_1 = \infty$, $u_2 = 0$ and $u_3 = 1$, and we introduced a step function

$$\Theta(u_2, \ldots, u_M) = \begin{cases} 1 & \text{for } u_2 > \cdots > u_M \\ 0 & \text{otherwise} \end{cases}. \tag{7.6}$$

It is with this formula that we will compute the 3-vertex of massless open string modes later. The computation will show, however, that for open strings on a single Dp-brane, the 3-vertex vanishes. So, to get something more interesting, we will first extend our formalism to stacks of N branes.

7.2 CHAN-PATON Construction

Once we consider a cluster of N Dp-branes on top of each other (i.e., in the same position), the string ends carry an effective color that encodes which brane the string end is actually attached to. Even before string theorists got interested in branes, colored string ends had been considered and implemented through the so-called CHAN-PATON construction [51]. With N possibilities for the color charge, each open string state can be characterized by one of the wave functions ψ together with an elementary matrix encoding our choice of colors:

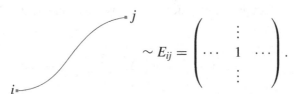

The only non-vanishing entry in the matrix E_{ij} is in the ith row and j^{th} column, corresponding to an open string with its left end on the ith brane and right end on the jth. The state space of such colored open bosonic string modes is therefore given by

$$\mathcal{H}^{OBS}_{N,Dp} = \mathcal{H}^{OBS}_{Dp} \otimes \text{Mat}(N \times N), \tag{7.7}$$

where we denoted the space of $N \times N$ matrices with $\text{Mat}(N \times N)$. Now we would like to pick any set of such wave functions from the space (7.7) and turn them into a number $\mathcal{A}$.

A state $|\psi^{ij}\rangle = |\psi; i, j\rangle$ with colors i and j now corresponds to the following matrix valued field:

$$\mathcal{O}_{\psi^{ij}} = \mathcal{O}_{\psi; i,j} = \int_{\mathbb{R}} du \, V_\psi(u) E_{ij}. \tag{7.8}$$

Any formula for a string amplitude must include a rule that encodes how numbers are extracted from the matrices. It is not hard to guess what this rule should be.

In fact, since colors cannot change along a boundary, the amplitude must be proportional to the trace over the color matrices of the external states:

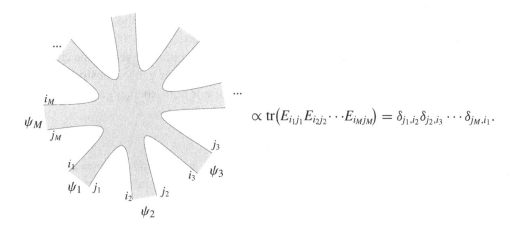

$$\propto \text{tr}\left(E_{i_1j_1}E_{i_2j_2}\cdots E_{i_Mj_M}\right) = \delta_{j_1,i_2}\delta_{j_2,i_3}\cdots\delta_{j_M,i_1}.$$

Hence, we conclude that taking the trace over the CHAN-PATON matrices is just what we need in order to ensure that non-zero amplitudes appear only for consistent choices of coloring. These comments may suffice to understand the following extension of our previous constructions to open string amplitudes on a stack of N identical branes:

$$\mathcal{A}(\psi_1^{i_1j_1},\ldots,\psi_M^{i_Mj_M})$$

$$\sim g_s^{-1}\int du_4\cdots du_M \sum_{\sigma\in S^{M-1}}\Theta\left(u_{\sigma(2)},\ldots,u_{\sigma(M)}\right)\text{tr}\left(E_{i_1j_1}E_{i_{\sigma(2)}j_{\sigma(2)}},\ldots,E_{i_{\sigma(M)}j_{\sigma(M)}}\right)$$

$$\cdot\left\langle V_1(\infty)V_{\sigma(2)}(u_{\sigma(2)})\cdots V_{\sigma(M)}(u_{\sigma(M)})\right\rangle. \quad (7.9)$$

The best way to digest such a complicated looking formula is to actually use it. We shall do that in the last part of this chapter. But in order to better understand the outcome of our string computations, we briefly digress to some simple field theory basics.

7.3 Non-Abelian Gauge Theory

We know that open strings on N Dp-branes give rise to vector fields with $p-1$ transverse degrees of freedom. This reminds us of a photon field A^μ, except that in our situation A^μ is matrix valued:

$$A_\mu(x) = A_\mu^{ij}(x)E_{ij}. \quad (7.10)$$

There exists a well-known generalization of the usual scalar valued vector bosons to matrix valued ones. Since matrices do not commute, a field theory involving matrix valued A_μ is known as non-Abelian gauge theory.

As the name suggests, theories of this kind are built such that they possess a symmetry under local gauge transformations with matrices $U = U(x)$. We imagine for a moment that the gauge field A^μ mediates the interaction between some charged fields ϕ with the simple behavior $\phi(x) \mapsto U(x)\phi(x)$ under gauge transformations. In order to find a sensible transformation law for our non-Abelian gauge field A^μ, we require that the covariant derivative

$$D_\mu \phi = (\partial_\mu - A_\mu)\phi \tag{7.11}$$

transforms in the same way as the field ϕ itself, i.e., we want $D^\mu(U\phi) = UD^\mu(\phi)$. Writing out this condition it is easy to determine the appropriate transformation law for our non-Abelian gauge field A^μ:

$$A_\mu \mapsto UA_\mu U^{-1} + (\partial_\mu U)U^{-1}. \tag{7.12}$$

In the case where A^μ and U are number valued, the first term is simply A^μ, and we recover the usual behavior of Abelian gauge fields under gauge transformations.

Next we would like to construct a field strength tensor with a simple behavior under gauge transformations. As one may easily check with the help of equation (7.12), the tensor

$$F_{\mu\nu} = \partial_\mu A_\nu - \partial_\nu A_\mu + \left[A_\mu, A_\nu\right] \tag{7.13}$$

transforms by conjugation with U:

$$F_{\mu\nu} \mapsto UF_{\mu\nu}U^{-1}. \tag{7.14}$$

In the Abelian case, the formula for the field strength simplifies and becomes the standard prescription to re-construct the electric and magnetic field from the vector potential. The latter are invariant under gauge transformations. The simple behavior of $F_{\mu\nu}$ under gauge transformations makes it straightforward to invent a gauge invariant action:

$$S[A] = \int d^D x \, \text{tr}\left(F_{\mu\nu}F^{\mu\nu}\right)$$

$$= 2\int d^D x \, \text{tr}\left(\partial_\mu A_\nu \partial^\mu A^\nu - \partial_\mu A_\nu \partial^\nu A^\mu\right) + \underbrace{4\int d^D x \, \text{tr}\left(\partial_\mu A_\nu \left[A^\nu, A^\mu\right]\right)}_{\text{3-vertex interaction}}$$

$$+ \underbrace{\int d^D x \, \text{tr}\left(\left[A_\mu, A_\nu\right]\left[A^\mu, A^\nu\right]\right)}_{\mathcal{O}(A^4)}.$$

$$\tag{7.15}$$

For an Abelian gauge field A^μ, only the quadratic terms remain in the action. This agrees with the fact that photons possess no self-interaction and is in contrast to the behavior of non-Abelian gauge bosons. Our aim now is to reproduce at least the third order terms from string theory. Since we will perform the string theory computation in momentum space, we note that the 3-vertex of non-Abelian gauge bosons reads

$$3\text{-vertex:} \quad \mathrm{tr}\left(k_1^\rho \zeta_\mu^1 \left[\zeta^{2\mu}, \zeta_\rho^3\right] + \cdots (\text{cyclic}) \cdots \right) \delta(k_1 + k_2 + k_3),$$

where ζ^a are three matrix valued polarizations, i.e. $\zeta_\mu^a = \xi_{\mu\,ij}^a E^{ij}$ and "(cyclic)" stands for two further terms that are obtained by cyclic permutation of the index $a = 1, 2, 3$.

7.4 The Open String 3-Point Vertex

Let us now turn to the long announced computation of the 3-point vertex of massless open strings. We recall from the previous chapters that the wave functions of such modes take the form

$$|\psi\rangle = \xi_\mu a_{-1}^\mu |k\rangle \tag{7.16}$$

with $k^2 = 0$ and transverse polarization $\xi_\mu k^\mu = 0$. To these, the state-field correspondence associates the following boundary vertex operators:

$$V_\psi(u) = \sqrt{\frac{2}{\alpha'}} : J^\mu(u) \exp(i k \cdot X(u)) : \xi_\mu, \tag{7.17}$$

where

$$J^\mu(u) = \sqrt{\frac{\alpha'}{2}} \sum_{n \in \mathbb{Z}} a_n^\mu u^{-n-1} \tag{7.18}$$

and $X^\mu(u)$ is obtained from the oscillators as reviewed in the first section. In order to find the scattering amplitude for the process depicted in figure 7.3 we have to compute the following field theory correlator:

$$\xi_\mu^1 \xi_\nu^2 \xi_\rho^3 \left\langle : J^\mu(u_1) V_{k_1}(u_1) :: J^\nu(u_2) V_{k_2}(u_2) :: J^\rho(u_3) V_{k_3}(u_3) : \right\rangle. \tag{7.19}$$

Let us comment that the computation of the correlator can certainly be performed without contracting the currents with the polarizations, but it turns out that the answer before such a contraction has a much more complicated dependence on the world-sheet coordinates u_a. The computation proceeds through the usual steps. First, we break the currents up into raising and lowering parts so that we can move the lowering parts to the right, where they annihilate the vacuum. Similarly, we move all raising parts to the left until all current insertions have been removed.

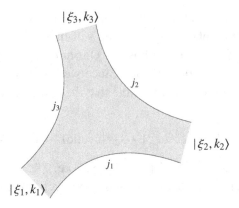

Figure 7.3 Three-point vertex of the open string.

The remaining 3-point function of exponential fields has been computed before, and it becomes rather trivial for three massless fields since $k_a k_b = 0$ for all $a, b = 1, 2, 3$. While moving currents around, they have to pass other currents or exponential fields. The corresponding commutators are given by

$$\left[J^\mu_\gtrless(u_1), V_k(u_2) \right] = \pm \frac{\alpha'}{2} \frac{k^\mu}{u_1 - u_2} V_k(u_2) \ , \quad \left[J^\mu_\gtrless(u_1), J^\nu(u_2) \right] = \pm \frac{\alpha'}{2} \frac{\eta^{\mu\nu}}{(u_1 - u_2)^2}.$$

(7.20)

The first relation contains one more current on the left-hand side than on the right and gives a factor $\alpha' k$ instead. Our second relation has an excess of two currents on the left-hand side, and only α' appears on the right. Consequently, there are two types of terms that contribute to the correlator: those that contain $(\alpha')^2$ and are linear in the momenta k_a and a second type of terms that are weighted with $(\alpha')^3$ and three momenta. The full answer is

$$\xi^1_\mu \xi^2_\nu \xi^3_\rho \left\langle : J^\mu(u_1) V_{k_1}(u_1) :: J^\nu(u_2) V_{k_2}(u_2) :: J^\rho(u_3) V_{k_3}(u_3) : \right\rangle$$

$$\sim \xi^1_\mu \xi^2_\nu \xi^3_\rho \left(\frac{\alpha'}{2} \right)^2 \frac{\eta^{\mu\nu} k^\rho_1 + \eta^{\mu\rho} k^\nu_3 + \eta^{\nu\rho} k^\mu_2 + \frac{\alpha'}{2} k^\mu_2 k^\nu_3 k^\rho_1}{(u_1 - u_2)(u_1 - u_3)(u_2 - u_3)} \delta(k_1 + k_2 + k_3).$$

(7.21)

In deriving the result we made frequent use of the transversality of the polarizations. Together with momentum conservation it implies $\xi^1 \cdot k_2 = -\xi^1 \cdot k_3$ etc. Such relations can be used to remove several terms with a more complicated dependence on the world-sheet coordinates. What remains is a rather simple function of the u_i.

Our prescription to put the three points to $u_1 = \infty, u_2 = 0, u_3 = 1$ removes the u-dependent denominator altogether.

At this point, we have almost completed the computation of the string amplitude. What remains is to sum over the two elements of the permutation group on two elements, i.e., we have to sum over the two orders 123 and 132. Let us denote the numerator of the above correlation function by C_{123}:

$$C_{123} = \left(\frac{\alpha'}{2}\right)^2 \left((\xi^1 \cdot \xi^2)(\xi^3 \cdot k_1) + \cdots (\text{cyclic}) \cdots \right.$$
$$\left. + \frac{\alpha'}{2}(\xi^1 \cdot k_2)(\xi^2 \cdot k_3)(\xi^3 \cdot k_1) \right) \delta(k_1 + k_2 + k_3).$$

Using once more momentum conservation and transversality of the polarizations, it is rather easy to check that C is antisymmetric with respect to an exchange of 2 and 3. Hence, if we sum over permutations of 2 and 3 before making the polarizations ξ^a matrix valued, then the corresponding 3-vertex of massless open strings is seen to vanish. We know this behavior from photons, and hence the massless vector particle on a single brane does indeed reproduce the 3-point vertex of a photon.

But the CHAN-PATON construction promotes all three polarizations ξ^a to $N \times N$ matrices $\zeta^a = \zeta^a_{ij} E^{ij}$. If we now sum over the two permutations, the sum no longer vanishes but rather gives

$$\mathcal{A}\left(\left| \zeta^1, k_1 \right\rangle \left| \zeta^2, k_2 \right\rangle \left| \zeta^3, k_3 \right\rangle \right)$$
$$= g_s^{-1} \sqrt{\alpha'} \, \text{tr} \left((k_1^\nu \zeta^1_\mu \left[\zeta^{2\mu}, \zeta^3_\nu \right] + \cdots (\text{cyclic}) \cdots + \frac{\alpha'}{2} k_1^\rho k_2^\mu k_3^\nu \zeta^1_\mu \left[\zeta^2_\nu, \zeta^3_\rho \right] \right)$$
$$\cdot \delta(k_1 + k_2 + k_3). \quad (7.22)$$

Apart from the familiar terms that we have seen in our discussion of non-Abelian gauge theory we also obtain a new term that does not appear for usual gauge bosons. Since it contains three momenta, the corresponding term in the space-time action would contain three derivatives. This term, however, is suppressed by the factor α'. Hence, to leading order in α', the scattering amplitude for the massless vector in the open string spectrum reproduced the 3-vertex of non-Abelian gauge theory, as we anticipated in the introduction. However, the usual gauge theory action receives stringy corrections that deform the particle theory as the strings get softer. Such string corrections to standard particle models certainly appear all over string theory, and they are indeed subject to extensive research.

Exercises

Problem 19. *Compute the 4-point amplitude for the scattering of four open string tachyons along the lines of the corresponding calculations for closed strings that was carried out in Section 4.4.*

HINT: *Use the integration formula*

$$\int_0^1 dy\, y^{a-1}(1-y)^{b-1} = B(a,b) = \frac{\Gamma(a)\Gamma(b)}{\Gamma(a+b)}.$$

8

Free Fermionic Quantum Field Theory

Throughout the preceding chapters we have used the bosonic string to gather some basic experience with a few key elements of string theory. But bosonic strings, both open and closed, had two central defects. To begin, all the particles we found in the spectrum had integer spin, i.e., there were no fermionic states to describe electrons, quarks, etc. The other problem was the existence of a tachyon among their vibrational modes. While both issues have not bothered us much in explaining some of the concepts and constructions of string theory, they certainly need to be addressed if we seriously hope to describe nature with string theoretic models.

In order to have fermionic states in the spectrum of string theory we will have to include fermionic fields into our 2-dimensional theory on the world-sheet. That will certainly not automatically ensure the absence of ghosts. The idea we shall follow to cope with tachyons is to aim for models with target space supersymmetry. Recall that supersymmetry guarantees that the M^2 is bounded from below by $M^2 \geq 0$. So, our strategy will be to include fermions on the world-sheet first. Then we shall combine the fermions with our bosonic fields X^μ into a model that possesses 2-dimensional (world-sheet) supersymmetry. Although we are ultimately looking for a string theory with target space rather than world-sheet supersymmetry, the results of this chapter will provide the crucial ingredient in succeeding with our principal aim.

8.1 Classical Fermionic Field Theory

As we anticipated, we shall begin by looking at a purely fermionic 2-dimensional free field theory. The action of the model is a functional depending on D real fermionic fields ψ^μ:

$$S[\psi] = -\frac{1}{4\pi} \int d\tau d\sigma \, \bar{\psi}^\mu \varrho^a \partial_a \psi^\nu \eta_{\mu\nu}. \qquad (8.1)$$

Here, we integrate again over a cylinder with MINKOWSKI metric and coordinates $\sigma \in [0, 2\pi]$ and $\tau \in \mathbb{R}$. The fields ψ^μ possess two real components,

$$\psi^\mu = \begin{pmatrix} \psi_-^\mu \\ \psi_+^\mu \end{pmatrix} \quad \text{and we defined } \bar{\psi}^\mu = (\psi^\mu)^t \varrho^0,$$

where the superscript t denotes the ordinary transpose. Furthermore, ϱ^0 and ϱ^1 are 2-dimensional DIRAC matrices, i.e. they satisfy the DIRAC algebra $\{\varrho^a, \varrho^b\} = 2\eta^{ab}$. An explicit representation is given by the following matrices:

$$\varrho^0 = \begin{pmatrix} 0 & -1 \\ 1 & 0 \end{pmatrix}, \quad \varrho^1 = \begin{pmatrix} 0 & 1 \\ 1 & 0 \end{pmatrix}. \tag{8.2}$$

As before, it is often useful to consider light-cone variables $\sigma^\pm = \tau \pm \sigma$ instead of σ and τ; see Section 2.3 for a detailed discussion of the coordinate transform. Inserting the explicit choice of matrices ϱ_a the action becomes

$$S[\psi] = \frac{1}{4\pi} \int d\sigma^- d\sigma^+ \eta_{\mu\nu} \left(\psi_-^\mu \partial_+ \psi_-^\nu + \psi_+^\mu \partial_- \psi_+^\nu \right). \tag{8.3}$$

In the case of the bosonic fields X^μ it was necessary to assume that X^μ depends periodically on the coordinate σ simply because X^μ was a parametrization of a surface in space-time. But for the fermionic fields the situation is different. In order to find consistent boundary conditions for fermions we shall take a look at the boundary terms in computing the variation of the action:

$$\delta S[\psi] = -\frac{1}{\pi} \int d\tau d\sigma \eta_{\mu\nu} \left(\partial_+ \psi_-^\mu \delta\psi_-^\nu + \partial_- \psi_+^\mu \delta\psi_+^\nu \right)$$
$$+ \frac{1}{4\pi} \int d\tau \eta_{\mu\nu} \left(\psi_-^\mu \delta\psi_-^\nu + \psi_+^\mu \delta\psi_+^\nu \right) \Big|_0^{2\pi}.$$

The first conclusion we can draw from this result concerns the equations of motions in the interior of the world-sheet. For the variation to vanish it is necessary that the fields $\psi_\pm^\mu$ satisfy

$$\partial_+ \psi_-^\mu = 0 \quad \text{and} \quad \partial_- \psi_+^\mu = 0, \tag{8.4}$$

i.e., the fields $\psi_\pm = \psi_\pm(\sigma^\pm)$ depend on a single light-cone coordinate. This is a behavior we have seen several times before, e.g., for the derivatives $\partial_\pm X^\mu$ or the VIRASORO fields $T_{\pm\pm}$. In addition to the equations of motion, we also see that $\psi_\pm$ must satisfy one of the following boundary conditions:

$$\psi_\mp(\sigma + 2\pi) = \begin{cases} \psi_\mp(\sigma) & \text{R-sector} \\ -\psi_\mp(\sigma) & \text{NS-sector} \end{cases}. \tag{8.5}$$

Here, R stands for RAMOND, and NS is short for NEVEU-SCHWARZ. Let us stress that the boundary conditions can be picked for left and right movers independently. This leaves us with four choices, denoted by RR, RNS, NSR, and NSNS.

From the action we can also compute the momentum $\Pi^\mu_\pm = \frac{1}{4\pi}\psi^\mu_\pm$ that is canonically conjugate to the fields $\psi^\mu_\pm$. This tells us that the equal time (anti-) POISSON brackets of our fermionic fields take the form

$$\left\{\psi^\mu_\mp(\tau,\sigma), \psi^\nu_\mp(\tau,\sigma')\right\}_{P.B.,+} = 4\pi\eta^{\mu\nu}\delta(\sigma - \sigma'). \tag{8.6}$$

Finally, let us also spell out the stress energy tensor of the theory. According to the usual rules, it is given by

$$T_{ab} = -\frac{1}{4}\left(\bar{\psi}^\mu\varrho_a\partial_b\psi_\mu + \bar{\psi}^\mu\varrho_b\partial_a\psi_\mu\right).$$

It is easy to see that T is symmetric, traceless, and conserved and hence, by the same reasoning we explained in the second chapter, it has only two non-vanishing components $T_{\pm\pm} = T_{\pm\pm}(\sigma^\pm)$, which each depend on a single light-cone coordinate, e.g.,

$$T_{--}(\sigma^-) = -\frac{1}{2}\psi^\mu_-\partial_-\psi^\nu_-\eta_{\mu\nu} = -\sum_{n=-\infty}^{\infty} l_n e^{-in\sigma^-}, \tag{8.7}$$

and similarly for T_{++}. On the right-hand side we expanded T_{--} in a FOURIER series. It is not difficult to verify that the FOURIER modes l_n obey the WITT-POISSON algebra, just as in the case of bosonic fields, i.e., $\{l_n, l_m\}_{P.B.} = (n - m)l_{n+m}$.

8.2 Quantization of the Fermionic Field

The quantization of the free fermionic theory is rather straightforward. Our presentation shall focus on the main new elements, in particular, the consequences of the various boundary conditions. The other important information we will take from this discussion is the value of the central charge for the VIRASORO algebra that arises from quantizing the l_n. To begin, let us expand the fields $\psi_\pm$ in terms of their FOURIER modes:

$$\psi^\mu_-(z) = \sqrt{-i}e^{-\frac{i}{2}\sigma^-}\psi^\mu_-(\sigma^-) = \begin{cases} \sum_{n\in\mathbb{Z}} b^\mu_n z^{-n-\frac{1}{2}} & \text{R-sector}, \\ \sum_{r\in\mathbb{Z}+\frac{1}{2}} b^\mu_r z^{-r-\frac{1}{2}} & \text{NS-sector} \end{cases}. \tag{8.8}$$

In the first line we have rewritten $\psi^\mu_-(\sigma^-)$ in terms of $z = \exp i\sigma^-$. Since the field ψ^μ has weight $h = 1/2$, an additional factor appears that is associated with the coordinate transformation. This factor also affects the periodicity properties of ψ_-^μ. Note, for example, that in the R-sector $\psi_-^\mu(z)$ now obeys $\psi_-^\mu(\exp(2\pi i)z) = -\psi_-^\mu(z)$. Similarly, we can proceed with the fields ψ_+^μ. Their modes are denoted $\bar{b}_n^\mu$ and $\bar{b}_r^\mu$. The anti-commutation relations of the modes can be worked out to take the form

$$\{b_n^\mu, b_m^\nu\} = \eta^{\mu\nu}\delta_{m+n,0} \quad \text{R-sector,} \tag{8.9}$$

$$\{b_r^\mu, b_s^\nu\} = \eta^{\mu\nu}\delta_{r+s,0} \quad \text{NS-sector.} \tag{8.10}$$

Let us now proceed to constructing the state space of the theory. This is particularly simple for the NS-sector, which has a unique ground state $|0\rangle$ such that

$$b_r^\mu|0\rangle = 0 \quad r > 0 \tag{8.11}$$

so that the space $\mathcal{H}_{\text{NS}}$ is spanned by states of the form

$$b_{r_1}^{\mu_1} \cdots b_{r_i}^{\mu_i}|0\rangle \quad \text{where} \quad r_1 < r_2 < \cdots < r_i \leq -\frac{1}{2}. \tag{8.12}$$

Since the fermionic modes b_r^μ obey $b_r^\mu b_r^\mu = 0$, a creation operator b_r^μ can be applied at most once. This explains why in the above basis we can restrict to $r_i < r_{i+1}$.

The R-sector is a bit more difficult to deal with since this time there are zero modes b_0^μ. These form the D-dimensional DIRAC algebra. For simplicity, let us assume that D is even (it will turn out to be $D = 10$ later). We will show momentarily that the zero modes possess a unique representation on a $2^{D/2}$-dimensional space of ground states. To see this, let us combine the zero modes into the following pairs of operators:

$$b_{00} = \frac{1}{\sqrt{2}}(b_0^1 - b_0^0), \qquad b_{00}^\dagger = \tfrac{1}{\sqrt{2}}(b_0^1 + b_0^0), \tag{8.13}$$

$$b_{0j} = \frac{1}{\sqrt{2}}(b_0^{2j+1} - ib_0^{2j}), \qquad b_{0j}^\dagger = \tfrac{1}{\sqrt{2}}(b_0^{2j+1} + ib_0^{2j}). \tag{8.14}$$

Note that each pair of directions in target space provides one pair of operators b and $b^\dagger$. It is easy to verify that the new operators behave like pairs of creation and annihilation operators:

$$\left\{b_{0i}, b_{0j}^\dagger\right\} = \delta_{ij}. \tag{8.15}$$

The space of ground states can therefore be generated with b_{0i} from a state $|R\rangle$ with the properties:

$$b_{0i}^\dagger|R\rangle = 0 \quad i = 0, \ldots, D/2 - 1, \qquad b_n^\mu|R\rangle = 0 \quad n > 0. \tag{8.16}$$

Since there are $D/2$ fermionic creation operators, the space of ground states has dimension $2^{D/2}$. From this space we can now generate the entire R-sector $\mathcal{H}_R$ by application of the raising operators b_n^μ with $n < 0$:

$$b_{n_1}^{\mu_1} \cdots b_{n_i}^{\mu_i} \prod_{l=0}^{\frac{D-2}{2}} b_{0l}^{\varepsilon_l} | R \rangle \quad n_1 < n_2 < \cdots < 0, \quad \varepsilon_l \in \{0,1\}. \tag{8.17}$$

Equations (8.11) and (8.12) define the state space of the NS-sector $\mathcal{H}_{NS}$ and eqs. (8.16) and (8.17) the state space of the R-sector $\mathcal{H}_R$. To these we have to add the left movers to obtain the full state space of the theory. It consists of the following four sectors:

$$\mathscr{H}_{RR}^{CFF} = \mathcal{H}_R \otimes \bar{\mathcal{H}}_R, \tag{8.18}$$

$$\mathscr{H}_{RNS}^{CFF} = \mathcal{H}_R \otimes \bar{\mathcal{H}}_{NS}, \tag{8.19}$$

$$\vdots$$

Following our usual conventions, we have placed a superscript CFF to refer to *closed fermionic field*. All these spaces carry a representation of the VIRASORO algebra. The VIRASORO generators read

$$L_m^R = \frac{1}{2} \sum_{n\in\mathbb{Z}} \left(n + \frac{m}{2}\right) : b_{-n}b_{m+n} : + \frac{D}{16}\delta_{m,0}^R, \tag{8.20}$$

$$L_m^{NS} = \frac{1}{2} \sum_{r\in\mathbb{Z}+\frac{1}{2}} \left(r + \frac{m}{2}\right) : b_{-r}b_{m+r} : . \tag{8.21}$$

The shift of L_0^R by a constant is necessary for the L_m^R to satisfy the usual VIRASORO commutation relations. With a bit of extra work one finds that the central charge of both L_n^R and L_m^{NS} is given by $c = D/2$, as opposed to $c = D$ in the case of free bosons.

8.3 Supersymmetric Free Field Theory

Let us now combine our new fermionic fields ψ^μ with the familiar bosonic fields X^μ into one action:

$$S[\psi, X] = -\frac{1}{4\pi} \int d\tau d\sigma \left(\frac{1}{\alpha'}\partial_a X^\mu \partial^a X_\mu + \bar{\psi}^\mu \varrho^a \partial_a \psi_\mu\right). \tag{8.22}$$

Being just a sum of the two models we studied before, the analysis of the new model and its quantization is trivially achieved by combining the results from above. But there is one new aspect that is not present in one of the two pieces. In fact, we shall show now that the action possesses a global supersymmetry. The supersymmetry transformations are given by

$$\delta X^\mu = \sqrt{\frac{\alpha'}{2}}\, \bar\epsilon \psi^\mu, \quad \delta \psi^\mu = \frac{1}{\sqrt{2\alpha'}}\, \varrho^a \partial_a X^\mu \epsilon \,. \tag{8.23}$$

Here ϵ is a two-component GRASSMANN valued constant. Variation of the action and insertion of the supersymmetry transformations (8.23) gives

$$\delta S[\psi, X] = \frac{1}{2\pi\sqrt{2}} \int d\tau d\sigma \left(\frac{1}{\alpha'} \partial_a \partial^a X^\mu \delta X_\mu - \delta\bar\psi^\mu \varrho^a \partial_a \psi_\mu \right)$$

$$= \frac{1}{2\pi\sqrt{2\alpha'}} \int d\tau d\sigma \left(\partial_a \partial^a X^\mu \bar\epsilon \psi_\mu - \epsilon^t \underbrace{(\varrho^a)^t \varrho^0 \varrho^b}_{=-\eta^{ab}\varrho^0} \partial_b \psi_\mu \partial_a X^\mu \right)$$

$$= \frac{1}{2\pi\sqrt{2\alpha'}} \int d\tau d\sigma \left(\partial_a \partial^a X^\mu \bar\epsilon \psi_\mu - \bar\epsilon \partial_a \partial^a X^\mu \psi_\mu \right) = 0\,. \tag{8.24}$$

Supersymmetry of the action implies that every field in the theory possesses a super-symmetric partner. This applies in particular to the symmetric, traceless, conserved stress energy tensor $T = T^F + T^B$ of our model. Its superpartner has the following form:

$$G_a := \frac{i}{\sqrt{2\alpha'}}\, \varrho^b \varrho_a \psi^\mu \partial_b X_\mu \quad \text{or} \quad G_\pm = i\sqrt{\frac{2}{\alpha'}} \psi^\mu_\pm (\sigma^\pm) \partial_\pm X_\mu\,. \tag{8.25}$$

In fact, if we act on G_a with the supersymmetry transformation (8.23), we recover the stress tensor; see problem 8.22. The current G_a has two important properties:

1. G_a is conserved, as one can see from the following simple computation that makes use of the equations of motion for both X^μ and ψ^μ:

$$\partial^a G_a = \frac{i}{\sqrt{2\alpha'}} \varrho^b \underbrace{\varrho_a \partial^a \psi^\mu}_{=0} \partial_b X_\mu + \frac{i}{\sqrt{2\alpha'}} \underbrace{\varrho^b \varrho^a}_{\eta^{ba}} \psi^\mu \partial_b \partial_a X_\mu = 0\,. \tag{8.26}$$

In the second term we have used that $\varrho^a \varrho_b$ is contracted with a symmetric tensor and that $\varrho^a \varrho^b + \varrho^b \varrho^a = 2\eta^{ab}$.

2. The current G_a satisfies $\varrho^a G_a = 0$. This is true simply because $\varrho^a \varrho^b \varrho_a = 0$.

The results are quite re-assuring. Recall that out of the four components that make up T, only two are independent. G_0 and G_1 have two components each. But the equations $\varrho^a G_a$ provide two equations that bring the number of independent components in G_a down to two, just as for the superpartner T. Moreover, the conservation of G_a implies that the light-cone components of G_a depend on one of the light-cone coordinates only, just as for the two independent components of the stress energy tensor, i.e., $G_- = G_-(\sigma^-)$ and $G_+ = G_+(\sigma^+)$.

8.4 Quantization and Superconformal Algebra

Before we end this chapter, let us now quantize our supersymmtric model. The state space of the quantized theory has four different sectors of the form

$$\mathcal{H}_{D,RR}^{CBFF} = \mathcal{H}_D^{CBF} \times \mathcal{H}_{RR}^{CFF}, \tag{8.27}$$

and similarly for RNS, NSR, and NSNS. All these spaces come equipped with the action of the VIRASORO modes $L_n = L_n^B + L_n^F$. Since the generators L_n^B in the bosonic theory commute with those in the fermionic part, the total central charge of the combined VIRASORO algebra is $c = D + D/2 = 3D/2$. But this is not the end of the story. The VIRASORO algebra is now extended by the modes of $G_\pm$ into an N=1 super-VIRASORO algebra. To spell out the relations of this new algebra, we transform the fields $G_\pm(\sigma^\pm)$ from $\sigma^\pm$ to z and $\bar{z}$ and then expand

$$G(z) = \begin{cases} \sum_{n\in\mathbb{Z}} G_n z^{-n-\frac{3}{2}} & \text{R-sector} \\ \sum_{r\in\mathbb{Z}+\frac{1}{2}} G_r z^{-r-\frac{1}{2}} & \text{NS-sector} \end{cases}. \tag{8.28}$$

The modes G_n and G_r may be expressed in terms of the modes a_n^μ and b_n^μ or b_r^μ of the fields ∂X^μ and ψ_-:

$$G_n = \sum_{m\in\mathbb{Z}} a_m^\mu b_{n-m}^\nu \eta_{\mu\nu} \quad \text{R-sector}, \tag{8.29}$$

$$G_r = \sum_{m\in\mathbb{Z}} a_m^\mu b_{r-m}^\nu \eta_{\mu\nu} \quad \text{NS-sector}. \tag{8.30}$$

From these formulas one can now determine the commutation relations of G_r and G_n with L_n and among the modes of G. The results read

$$[L_m, G_r] = \left(\frac{1}{2}m - r\right) G_{m+r}, \tag{8.31}$$

$$\{G_r, G_s\} = 2L_{r+s} + \frac{D}{2}\left(r^2 - \frac{1}{4}\right)\delta_{r+s,0} \tag{8.32}$$

in the NS-sector and the same for the R-sector with $r \to n, s \to m$. The extended N=1 super-VIRASORO algebra will later provide the constraints that we need to implement in order to descend from the state space of the field theory to the state space of superstrings.

Exercises

Problem 20. *Determine the 2-point function of a real fermionic field (8.8) in the ground state $|0\rangle$ of the* NEVEU-SCHWARZ *sector, i.e., in a state satisfying $b_r|0\rangle = 0$ for all $r \geq 1/2$.*

Problem 21. *Compute the expectation value of a product of two real fermionic fields in the ground state $|R\rangle$ of the* RAMOND *sector. Compare the result to the outcome of the previous exercise.*

Problem 22. *Determine the superpartner δG_a of the field $G = (G_a)$ defined in eq. (8.25) using the supersymmetry transformations (8.23).*

Problem 23. *Consider a set of fermionic operators b_n^μ that satisfy the anti-commutation relations (8.9) along with our usual bosonic oscillators a_n^μ with commutation relations (3.2). Define operators G_n through eq. (8.29) and show that they obey the commutation relations (8.32) with $L_n = L_n^b + L_n^R$. Explicit formulas for L_n^b and L_n^R were spelled out in eqs. (3.27) and (8.20), respectively.*

9

Supersymmetry in 10 Dimensions

We are just about to construct the spectrum of type IIA/B superstring theory. But before we get there, we have to pause for a moment and learn a bit about the 10-dimensional supersymmetry that organizes the spectrum of these models. The supersymmetry algebra is a Lie superalgebra in which the usual POINCARÉ algebra is extended by fermionic generators Q_a. The latter transform in one or several spinor representations of the LORENTZ algebra. In order to spell out the construction of superalgebras and to investigate their multiplets, we need a bit of background on representations of the POINCARÉ algebra. In particular, we will have to study the finite dimensional spinor representations. We provide the necessary background in the first two sections before spelling out the defining relations of the N=1 and N=2 superalgebras in $D = 10$ dimensions. Understanding the structure of their massless multiplets is the principal goal of this chapter. While most of the methods we shall describe below can be used in dimensions other than $D = 10$, some of the results are rather dependent on the choice of dimension. So, we shall restrict ourselves to the 10-dimensional case, the one that will turn out to be relevant for superstring theory. While the following discussion of supersymmetry should be sufficient for most of the later chapters, it is certainly helpful to consult one of the many textbooks, such as [67, 73, 75], for further background.

9.1 The POINCARÉ Algebra and Its Representations

Symmetries of the 10-dimensional MINKOWSKI space-time are described by the POINCARÉ algebra $so(1, 9) \ltimes \mathbb{R}^{1,9}$ with its generators $M^{\mu\nu}$, P^{μ} satisfying the following commutation-relations:

$$\left[M^{\mu\nu}, M^{\rho\sigma}\right] = i\eta^{\mu\rho}M^{\nu\sigma} - i\eta^{\mu\sigma}M^{\nu\rho} + i\eta^{\nu\sigma}M^{\mu\rho} - i\eta^{\nu\rho}M^{\mu\sigma}, \qquad (9.1)$$

$$\left[M^{\mu\nu}, P^{\rho}\right] = i\eta^{\mu\rho}P^{\nu} - i\eta^{\nu\rho}P^{\mu}. \qquad (9.2)$$

The space of one-particle wave functions of any relativistic 10-dimensional theory may be decomposed into irreducible representations (multiplets) of the POINCARÉ algebra. It will be useful to have a mathematically precise understanding about how to construct these representations. The construction proceeds in several steps.

Step 1: Choose an orbit $\mathcal{O}$ of so(1,9) action on $\mathbb{R}^{1,9}$. The two types of orbits that will be relevant for us are

$$\mathcal{O}_* = \{0\} \quad \text{and} \quad \mathcal{O}_{m^2} = \{ p^2 = m^2 \}. \tag{9.3}$$

These orbits contain all values of the momentum that can be measured for a given multiplet. The orbit $\mathcal{O}_{m^2}$, for example, contains all the momenta for a multiplet of a particle with mass m. Only multiplets with $m^2 \geq 0$ will play a role below. Those with $m^2 < 0$ are tachyonic. The orbit $\mathcal{O}_*$ does not give rise to a particle multiplet, but it will still lead to an important class of representations.

Before we continue, let us pick one particular point $p_0 = (p_0^\mu)$ in the orbit $\mathcal{O}$. The standard choices are $p_0^* = 0, p_0(p^2 = 0) = (1,1,0,\ldots,0)$ and $p_0(p^2 = m^2) = (m,0,\ldots,0)$. The representations we are about to construct will not depend on this choice.

Step 2: Next we are instructed to determine the so-called stabilizer subalgebra of the point p_0, i.e., the subalgebra of all LORENTZ boosts and rotations $M^{\mu\nu}$ that leave the point p_0 invariant. This algebra depends only on the orbit of p_0. Different choices of p_0 on the same orbit lead to isomorphic stabilizer subalgebras. For the orbits we listed above, the stabilizer subalgebras are

$$\mathfrak{g}_* = \text{so}(1,9), \tag{9.4}$$
$$\mathfrak{g}_0 = \text{iso}(8) \cong e(8) \quad \text{for } m^2 = 0, \tag{9.5}$$
$$\mathfrak{g}_{m^2} = \text{so}(9) \quad \quad \text{for } m^2 > 0. \tag{9.6}$$

The massless case requires some additional comments. ISO(8) denotes the group of isometries of $\mathbb{R}^8$. Another common name is *Euclidean group* E(8). This group is generated by translations and rotations of an 8-dimensional Euclidean space. Its appearance as a stabilizer subgroup of massless orbits is reviewed, e.g., in [72].

In our second step we are now instructed to choose an irreducible unitary representation of stabilizer subalgebra $\mathfrak{g}_\mathcal{O}$ on some finite dimensional vector space V. For the orbits $\mathcal{O}_{m^2}$ the choice of V is associated with the spin of the corresponding particle multiplet. Let us note that the only finite dimensional unitary representations of the Euclidean algebra e(n) are those in which the translations t(8) are trivial, and hence we must simply choose a finite dimensional representation of the rotation algebra so(8) = e(8)/t(8). In conclusion, the second step of the program requires good command of the representation theory of so(1,9), so(8), and so(9). We shall see a bit more of that below.

Step 3: The data we have chosen so far give rise to an action of the subalgebra $\mathfrak{g}_{\mathcal{O}} \ltimes \mathbb{R}^{1,9}$ on the space V if we declare P^{μ} to act as multiplication operators with the constants $p_0^{\mathcal{O}}$. From here we proceed to a representation of the full POINCARÉ algebra by a process called *induction*. The idea is to formally add states that are obtained from V by the action of boosts. Unless we start with $\mathcal{O} = \mathcal{O}_*$, this leads to an infinite dimensional representation of the POINCARÉ algebra on the space

$$\mathcal{H}_{\mathcal{O}}^V := \left(\mathrm{so}(1,9) \ltimes \mathbb{R}^{1,9} \right) \otimes_{(\mathfrak{g}_{\mathcal{O}} \ltimes \mathbb{R}^{1,9})} V \cong L^2(\mathcal{O}; V) . \tag{9.7}$$

The symbols defining $\mathcal{H}_{\mathcal{O}}^V$ might seem a bit frightening at first. What they mean is that we should start from the usual tensor product of the POINCARÉ algebra with the space V. This space has an obvious action of the POINCARÉ generators that leaves V untouched, but it is certainly not irreducible. The subscript on the tensor product $\otimes$ instructs us to take all elements from the subalgebra $\mathfrak{g}_{\mathcal{O}} \ltimes \mathbb{R}^{1,9}$ out of the first factor and to let them act on their representation space V. One can realize elements of the resulting space as functions on the orbit $\mathcal{O}$ with values in the vector space V. In this way we have now constructed a representation of the full POINCARÉ algebra; see, e.g., [72] for more details.

Claim: The construction outlined in the previous three steps gives unitary irreducible representations of the POINCARÉ algebra, and all unitary irreducible representations are of this form. The finite dimensional unitary irreducible representations of the POINCARÉ algebra are obtained with $\mathcal{O}_*$. In other words, as soon as the momentum operators are represented non-trivially, we end up with an infinite dimensional representation of the POINCARÉ algebra. This conclusion is certainly not surprising.

Examples: If we pick $\mathcal{O} = \mathcal{O}_{m^2}$ and $V = \mathbb{R}$, the trivial representation of the stabilizer algebra $\mathfrak{g}_{\mathcal{O}}$, we obtain massless and massive scalar multiplets. For $\mathcal{O} = \mathcal{O}_0$ and $V = \mathbf{8}_V$, the 8-dimensional vector representation of so(8) $= \mathfrak{g}_{\mathcal{O}}$, we end up with a massless vector multiplet. Massive vector multiplets are obtained from $\mathcal{O} = \mathcal{O}_{m^2}$ and the 9-dimensional vector representation $V = \mathbf{9}_V$ of so(9). We see that such multiplets contain one additional polarization.

9.2 Spinor Representation of so(1, 9)

We are now going to construct some important finite dimensional representations of the POINCARÉ algebra. As we reviewed in the previous discussion, in any finite dimensional representation of the POINCARÉ algebra, the momenta P^μ must be represented by zero so that we are effectively dealing with a representation of the LORENTZ algebra so(1, 9). Before we go into the details of the construction let us stress that all the ideas we sketch below apply to spinor representations of the LORENTZ algebra in any (even) dimension D. The results, however, depend very much on the choice of D. We shall therefore restrict our presentation to $D = 10$, the case that will turn out to be relevant for the superstring, and comment on a few other dimensions only here and there.

The basic idea is to construct representations of the LORENTZ algebra from a representation of the DIRAC algebra

$$\{\gamma^\mu, \gamma^\nu\} = 2\eta^{\mu\nu} \tag{9.8}$$

through the following formula:

$$\Sigma^{\mu\nu} = -\frac{i}{4}[\gamma^\mu, \gamma^\nu]. \tag{9.9}$$

We have learned in the previous chapter how to build a representation of the DIRAC algebra. In fact, the DIRAC algebra appeared there as the algebra of fermionic zero modes in the RAMOND sector in the free fermionic field theory. The idea was to combine pairs of dimensions and to form one creation and one annihilation operator for each pair. In a 10-dimensional theory this gives five creation operators that generate a 32-dimensional representation space S. Being linear combinations of the creation and annihilation operators, the generators γ^μ of the DIRAC algebra act on S. Hence, we can represent the LORENTZ generators $\Sigma^{\mu\nu}$ as 32-dimensional matrices.

The representation of the LORENTZ algebra so(1, 9) turns out to be reducible. As we shall see momentarily, the space S decomposes into two 16-dimensional subspaces $S_\pm$ that carry an irreducible action of the $\Sigma^{\mu\nu}$. The projection to the irreducible subspaces can be constructed out of the following 10-dimensional analogue γ^{11} of the matrix γ^5 that we know from 4-dimensional fermions:

$$\gamma = \gamma^{11} := \gamma^0\gamma^1\gamma^2\cdots\gamma^9. \tag{9.10}$$

As one may check, for example, by explicit construction, the matrix γ^{11} has the following important properties:

1. The matrix γ^{11} anti-commutes with DIRAC matrices γ^μ:

$$\{\gamma^{11}, \gamma^\mu\} = 0.$$

Since all the involved matrices can be constructed explicitly, it is in principle straightforward to check this relation.

2. From the previous property of the matrix γ^{11} and the construction of the LORENTZ generators (9.9) it is now easy to conclude that γ^{11} commutes with $\Sigma^{\mu\nu}$:

$$\left[\gamma^{11}, \Sigma^{\mu\nu}\right] = 0.$$

By SCHUR's lemma, the representation of so(1,9) on S is therefore reducible if B is not proportional to the identity matrix (see below).

3. The matrix $\gamma = \gamma^{11}$ squares to the identity matrix:

$$\gamma^2 = \mathbf{1}_{32\times32}.$$

This means that its eigenvalues can assume only the values ±1. Closer inspection shows that each of them appears 16 times.

From all this we conclude that the representation space S decomposes into two inequivalent representations $S_\pm$ of so(1,9),

$$S = S_+ \oplus S_- \quad \text{with} \quad S_\pm = \Pi_\pm S, \tag{9.11}$$

where $\Pi_\pm$ are the following projection operators:

$$\Pi_\pm = \frac{1}{2}\left(1 \pm \gamma^{11}\right). \tag{9.12}$$

In this way we have now decomposed the 32-dimensional DIRAC representation of the LORENTZ algebra into two 16-dimensional so-called WEYL representations $S_\pm$. The latter are actually irreducible. Similar decompositions work in all even dimensions. In $D = 4$ dimensions, for example, the DIRAC representation $S^{(4)}$ has dimension $\dim S^{(4)} = 2^{D/2} = 4$, and it splits into two 2-dimensional WEYL multiplets $S_\pm^{(4)}$. Similarly, the 2-dimensional DIRAC multiplet $S^{(2)}$ in a 2-dimensional space-time decomposes into two 1-dimensional WEYL representations, a fact we already had seen in the previous chapters.

When we discussed fermions on the 2-dimensional world-sheet, we had also assumed that the two WEYL-spinors $\psi_\pm$ were real. Such reality properties are strongly dependent on the dimension. We shall now discuss how to investigate this issue more systematically, based on the properties of a second auxiliary matrix B that is defined by

$$B = \gamma\gamma^3\gamma^5\gamma^7\gamma^9.$$

Note that the matrix γ^1 is not included in this product. The 32-dimensional matrix B has a few properties that relate it to the reality properties of our representations:

1. The complex conjugate $(\Sigma^{\mu\nu})^*$ of the matrix elements of the LORENTZ generators satisfies

$$(\Sigma^{\mu\nu})^* = -B\Sigma^{\mu\nu}B^{-1}.$$

We stress that the symbol $*$ in this equation *does not* involve any transpose. If B
was the trivial identity matrix, the previous relation would imply that the matrix
elements $\Sigma^{\mu\nu}$ were purely imaginary. The matrix B, however, is not trivial, and
hence it controls the failure of $\Sigma^{\mu\nu}$ to have imaginary matrix elements.

2. The matrix B commutes with the matrix γ^{11} we have introduced in our discussion of WEYL multiplets:

$$B\gamma^{11}B^{-1} = \gamma^{11}.$$

Hence, B acts within the two 16-dimensional eigenspaces of the matrix γ^{11}. This
property depends very much on the dimension D. In 4-dimensional MINKOWSKI
space, for example, the analogue of B maps the two eigenspaces of γ^5 onto each
other.

3. The matrix B satisfies the following reality property:

$$B^*B = 1_{32\times32}.$$

Recall that $*$ simply denotes complex conjugation of matrix elements.

We now apply these properties of B to construct the real 32-dimensional vector
space

$$S^R = \left\{\psi^* = B\psi \mid \psi \in S\right\} \tag{9.13}$$

of all elements ψ in S that have real components up to multiplication by the matrix
B. The definition of S^R is consistent since

$$(\psi^*)^* = (B\psi)^* = B^*\psi^* = B^*B\psi = \psi.$$

In the last step we have used the third property of B: that $B^*B = 1$. The space S^R
has been built such that it admits an action of the generators $i\Sigma^{\mu\nu}$:

$$(i\Sigma^{\mu\nu}\psi)^* = -i(\Sigma^{\mu\nu})^*\psi^* = iB\Sigma^{\mu\nu}B^{-1}B\psi = B(i\Sigma^{\mu\nu}\psi).$$

Note that elements of the LORENTZ group are obtained by exponentiating the generators $i\Sigma^{\mu\nu}$. Our short calculation made crucial use of the twisted reality properties
of $\Sigma^{\mu\nu}$. What we have shown so far is that we can impose an appropriate reality
condition on the DIRAC representation of the LORENTZ algebra. We call the corresponding representation on S^R a MAJORANA representation of so$(1,9)$. Again,
MAJORANA representations exist in all even dimensions.

Finally, because the matrix B commutes with the matrix γ^{11}, we can conclude that, in 10-dimensional MINKOWSKI space, our split of the DIRAC representation into two WEYL multiplets is consistent with the reality condition we imposed. In other words, we can split S^R into two 16-dimensional real subspaces:

$$S^R = S^R_+ \oplus S^R_- \quad \text{where} \quad S^R_\pm = \Pi_\pm S^R.$$

The spaces $S^R_\pm$ carry actions of the LORENTZ algebra. We call these representations MAJORANA-WEYL since they satisfy WEYL- and MAJORANA-conditions simultaneously. Let us add that MAJORANA-WEYL multiplets exist in dimensions $D = 2 + 4n$. In this respect, the space-time dimensions $D = 2$ and $D = 10$ are a bit special.

In summary, we have now shown that there exist two 16-dimensional real spinor representations of the LORENTZ algebra so$(1,9)$. So far, we have denoted the inequivalent MAJORANA-WEYL multiplets by $S^R_\pm$. Below we shall often use the simpler notation $S^R_+ = \mathbf{16}$ and $S^R_- = \mathbf{16}'$ that is quite common in the physics literature.

There is one final remark we would like to add before we proceed to the discussion of superalgebras. It concerns the decomposition of the MAJORANA-WEYL multiplets $\mathbf{16}$ and $\mathbf{16}'$ into representations of so$(8) \subset$ so$(1,9)$. This requires a bit of background on representations of so(8). All we need to know here is that the LIE algebra so(8) possesses three inequivalent 8-dimensional representations. One of them is the fundamental (or vector) representation, which is often denoted by $\mathbf{8}_V$. But in addition there are two more 8-dimensional spinor representations. They are essentially constructed in the same way in which we constructed our spinor representations of the LORENTZ algebra. One departs from the corresponding version of the DIRAC algebra in which the metric is Euclidean and pairs directions to form four creation operators. The resulting representation of so(8) is 16-dimensional. As before, it turns out to be reducible and to decompose into two 8-dimensional spinor representations denoted by $\mathbf{8}_S$ and $\mathbf{8}_C$. After this preparation we can now spell out how the two 16-dimensional representations $\mathbf{16}$ and $\mathbf{16}'$ of so$(1,9)$ decompose with respect to the action of so$(8) \subset$ so$(1,9)$:

$$\mathbf{16} = \mathbf{8}_V \oplus \mathbf{8}_S, \qquad \mathbf{16}' = \mathbf{8}_V \oplus \mathbf{8}_C. \tag{9.14}$$

We shall need these two decomposition formulas later in our discussion of massless supermultiplets.

9.3 10-Dimensional Superalgebras

We are finally well prepared to construct the superalgebras that are relevant for 10-dimensional superstring theory. The construction departs from the POINCARÉ algebra that is generated by $M^{\mu\nu}$ and P^{μ}. We would like to add some fermionic generators Q such that the new algebraic structure closes under taking (anti-) commutators. To this end, we organize the fermionic generators into finite dimensional multiplets of so(1, 9):

$$[P^{\mu}, Q_a] = 0, \qquad [M^{\mu\nu}, Q_a] = (\Sigma^{\mu\nu})_a^b Q_b. \qquad (9.15)$$

For the right-hand side of the second commutator to make sense the index a should label basis elements in a 16-dimensional WEYL representation of the LORENTZ algebra. We conclude that fermionic generators must be added in multiples of 16. Here we shall discuss adding either one or two such multiplets. The associated superalgebras are known as N=1 and N=2 POINCARÉ superalgebra, respectively.

N = 1 POINCARÉ *superalgebra.*

The N=1 POINCARÉ superalgebra is the minimal extension of the POINCARÉ algebra in which we add 16 supercharges Q_a in a single WEYL representation 16 or 16$'$ of the LORENTZ algebra. We have displayed the commutation relations between the Q_a and the POINCARÉ generators already. So, it remains to spell out the anti-commutators of the supercharges themselves. In the N=1 POINCARÉ superalgebra these take the form

$$\{Q_a, \overline{Q}_b\} = -2P_{\mu} (\Pi_{\pm}\gamma^{\mu})_{ab} \ . \qquad (9.16)$$

Here, we have inserted the projectors $\Pi_{\pm}$ to associate a 16×16 matrix $\Pi_{\pm}\gamma^{\mu} = \Pi_{\pm}^2\gamma^{\mu} = -\Pi_{\pm}\gamma^{\mu}\Pi_{\mp}$ to the DIRAC matrix γ^{μ}. In fact, our short calculation in the last sentence showed that $\Pi_{\pm}\gamma^{\mu}$ acts trivially on the subspace $S_{\pm}$ and hence that it is completely characterized by its action on the 16-dimensional subspace $S_{\mp}$. In spelling out the anti-commutator of supercharges Q_a it was convenient to introduce the short-hand $\overline{Q}_a$ for the following linear combination of supercharges:

$$\overline{Q} = Q^T C = Q^T B \gamma^0. \qquad (9.17)$$

The charge conjugation matrix C is defined as a product $C = B\gamma^0$. It will also feature in other discussions below. This concludes our construction of the N=1 POINCARÉ superalgebra.

N = 2 POINCARÉ *superalgebra.*

We now want to study extensions of the POINCARÉ algebra by two MAJORANA-WEYL multiplets of supercharges. In other words, we would like to adjoin supercharges Q_a^A with $a = 1, \ldots, 16$ and $A = 1, 2$. When we fix the transformation properties of the two 16-dimensional multiplets Q^1 and Q^2 by imposing commutation relations with the $M^{\mu\nu}$ we have the choice between the two representations **16** and **16**$'$ of the LORENTZ algebra. This results in a total of four different possibilities:

$$\mathbf{16} \oplus \mathbf{16} \text{ or } \mathbf{16}' \oplus \mathbf{16}' \ \rightarrow \ \text{type IIB},\tag{9.18}$$

$$\mathbf{16} \oplus \mathbf{16}' \text{ or } \mathbf{16}' \oplus \mathbf{16} \ \rightarrow \ \text{type IIA}.\tag{9.19}$$

The choice in the first line gives rise to what is known as type IIB POINCARÉ superalgebra. Similarly, one calls a POINCARÉ superalgebra type IIA if the supercharges transform in the two inequivalent MAJORANA-WEYL representations of so(1,9). Let us turn to type IIB first. In this case, the relations between supercharges read

$$\left\{ Q_a^A, \overline{Q}_b^B \right\} = -2\delta^{AB} P_\mu \, (\Pi_\pm \gamma^\mu)_{ab} \, .\tag{9.20}$$

Our notations are the same as in the discussion of the N=1 POINCARÉ superalgebra. In the case of type IIA, we have

$$\left\{ Q_a^1, \overline{Q}_b^1 \right\} = -2P_\mu \, (\Pi_+ \gamma^\mu)_{ab} \, ,\tag{9.21}$$

$$\left\{ Q_a^2, \overline{Q}_b^2 \right\} = -2P_\mu \, (\Pi_- \gamma^\mu)_{ab} \, .\tag{9.22}$$

Thereby we have constructed the three 10-dimensional superalgebras that will play a role in superstring theory. As we shall discuss in the next chapter type IIA/B POINCARÉ superalgebras describe the space-time symmetries of type IIA/B superstring theory. The N=1 POINCARÉ superalgebra, on the other hand, appears for the heterotic string.

N=2 Massless Supermultiplets

We consider the **type IIB** case first. We are looking for the massless representation of the N=2 POINCARÉ superalgebra. These are constructed using the same ideas we spelled in the first section. Note that the supercharges act trivially on the momenta. Hence, we modify our constructions such that V now includes an action of the following algebra:

$$\left\{ Q_a^A, \overline{Q}_b^B \right\} = -2\delta^{AB} (p_0)_\mu \, (\Pi_\pm \gamma^\mu)_{ab} \, ,\tag{9.23}$$

where we have simply replaced the momentum operators P_μ by their constant values on V. After our induction, the POINCARÉ superalgebra is then represented on the space $L_2(\{p^2 = 0\}|V)$ of square integrable functions that take values in V. The space V still carries an action of so(8), but it now decomposes into several irreducible representations that are connected only by the action of the supercharges.

In order to understand the massless supermultiplets of the type IIB algebra we need a bit of additional input. Recall that we begin with 32 supercharges. On V, these obey the relations (9.23), i.e. their anti-commutators are just numbers. Closer inspection of their algebra shows that 16 of the supercharges possess trivial anti-commutators, while the remaining 16 can be combined into two sets of eight creation and annihilation operators. Under the action of so(8) the anti-commuting charges transform in the two 8-dimensional vector representations $\mathbf{8}_V$. The creation and annihilation operators, on the other hand, form two multiplets $\mathbf{8}_S$ (or $\mathbf{8}_C$, depending on the choice we made when constructing our type IIB superalgebra). It is consistent to let the 16 anti-commuting supercharges act trivially on V. Hence, we shall build V simply from some so(8) invariant "ground" state by the action of 8 fermionic creation operators. Under the action of so(8), the resulting space V decomposes into the following representations:

$$V \cong \underbrace{\mathbf{1}}_{\text{dilaton}} \oplus \underbrace{\mathbf{8}_S}_{\text{dilatino}} \oplus \underbrace{\mathbf{28}}_{\text{KR-field}} \oplus \underbrace{\mathbf{56}_S}_{\text{gravitino}} \oplus \underbrace{\mathbf{70}}_{\substack{\text{graviton} \\ 4^* - \text{form}}} \oplus \underbrace{\mathbf{56}_S}_{\text{gravitino}} \oplus \underbrace{\mathbf{28}}_{2-\text{form}} \oplus \underbrace{\mathbf{8}_S}_{\text{dilatino}} \oplus \underbrace{\mathbf{1}}_{0-\text{form}} .$$

$$(9.24)$$

The numbers $1, 8, 28$, etc. count the number of states one can build from the ground states by acting with $0, 1, 2, \ldots$ fermionic creation operators. Except for the 70-dimensional space that is built with four creation operators, all others carry an irreducible representation of so(8). We have given them names already, corresponding to how they are realized in type IIB supergravity.

Let us briefly go through the bosonic multiplets. The dilaton Φ is a scalar and so has a single component. The next entry $\mathbf{8}_S$ is obtained by applying a single fermionic creation operator, and hence it is fermionic. So we proceed to the $\mathbf{28}$ right away. It is the KALB-RAMOND field B. As a massless antisymmetric tensor field B indeed has $8 \cdot 7/2 = 28$ components. The next bosonic entry is the 70. This is a bit more subtle because $\mathbf{70}$ is not the dimension of an irreducible representation of so(8). It actually decomposes into two irreducible representations of dimension 35 each. One of these two multiplets corresponds to a massless graviton g, i.e. a traceless, symmetric tensor with $8 \cdot 7/2 + 7 = 35$ components. The other 35 count the degrees of freedom of a self-dual 4-form $C^4 = *C^4$ where $*$ denotes the HODGE dual. Note that an ordinary 4-form would have 70 components. By the self-duality condition we remove 35 of them. Going further on our list we find another representation, $\mathbf{28}$, which we now attribute to a 2-form field C^2. At the very end, there appears another bosonic scalar multiplet. We associate it with a 0-form field C^0. In summary, our supermultiplet contains 128 bosonic components forming the dilaton, KALB-RAMOND field, graviton, and a number of form fields C^0, C^2, C^4.

In addition there are four fermionic multiplets. These include the dilatinos and gravitinos.

One may perform a similar analysis for the **type IIA** case. Again 16 of the 32 supercharges act trivially on V, which is then built with eight creation operators from some invariant ground state. Hence, the massless type IIA supermultiplet has the same number of components as in the case of type IIB. What makes the analysis a little more difficult is that these creation operators do not have a definite transformation law under the so(8). So, one has to recombine the various subspaces into representations of so(8) by hand. The result is given by

$$V \cong \underbrace{\mathbf{1}}_{\text{dilaton}} \oplus \underbrace{\mathbf{8}_V}_{\text{vector}} \oplus \underbrace{\mathbf{28}}_{\text{KR-field}} \oplus \underbrace{\mathbf{35}_V}_{\text{graviton}} \oplus \underbrace{\mathbf{56}_V}_{\text{3-form}} \oplus \underbrace{\mathbf{8}_S \oplus \mathbf{8}_C \oplus \mathbf{56}_S \oplus \mathbf{58}_C}_{\text{non-chiral fermionic}} . \qquad (9.25)$$

The first five irreducible representations are bosonic and have 128 components, as before. In addition to the familiar dilatonΦ, KALB-RAMOND field B, and graviton g there is a vector multiplet that may be represented by a 1-form C^1 and a 3-form field C^3. The last four terms describe four fermionic multiplets with a total number of 128 components.

Exercises

Problem 24. *Construct a matrix representation of the 2-dimensional Dirac algebra*

$$\{\gamma^\mu, \gamma^\nu\} = 2\eta^{\mu\nu}$$

in the basis consisting of $e_1 = |0\rangle$ and $e_2 = b|0\rangle$ where $|0\rangle$ satisfies $b^\dagger|0\rangle = 0$ and $b = (\gamma^1 - \gamma^0)/2, b^\dagger = (\gamma^1 + \gamma^0)/2$.

(b) Show that $\gamma = \gamma^0\gamma^1$ commutes with $\Sigma^{\mu\nu} = -i[\gamma^\mu, \gamma^\nu]/4$ and squares to the identity matrix. Determine the subspaces that carry the two Weyl multiplets.

(c) Show that $B = \gamma$ satisfies $(\Sigma^{\mu\nu})^* = -B\Sigma^{\mu\nu}B^{-1}$ as well as $B\gamma = \gamma B$ and $B^*B = 1$. Determine the real subspaces that carry the MAJORANA-WEYL multiplets.

Problem 25. *Construct an irreducible representation of the Dirac algebra in $D = 4$ dimensions. This allows to build a representation of the Lorentz algebra through the prescription*

$$\Sigma^{\mu\nu} = -\frac{i}{4}[\gamma^\mu, \gamma^\nu].$$

(a) Show that $\gamma^5 = \kappa\gamma^0\gamma^1\gamma^2\gamma^3$ is real with an appropriate choice of the constant κ and that it commutes with the Lorentz generators.

(b) Construct the matrix $B = \kappa'\gamma^5\gamma^3$ with κ' chosen such that $B^*B = 1$. Demonstrate that

$$(\Sigma^{\mu\nu})^* = -B\Sigma^{\mu\nu}B^{-1}, \quad B\gamma^5 = -\gamma^5 B.$$

From the results answer the question whether there are MAJORANA-WEYL *fermions in four dimensions.*

Problem 26. *The N=1* POINCARÉ *superalgebra in four dimensions is generated by the bosonic generators $M^{\mu\nu}, P^\mu$ and four additional fermionic generators Q_a that transform in a* MAJORANA *representation of so(1,3). Their anti-commutation relations are of the form*

$$\{Q_a, \overline{Q}_b\} = -2P_\mu\gamma^\mu_{ab}$$

with $\overline{Q}_b = Q_a(B\gamma^0)^a_b$. Construct representations with $P^2 = 0$ and with half-integer spin $s \leq 2$.

10

Construction of Type IIA/B Superstrings

We have reached a decisive moment in these chapters. Let us recall our main motivation to go beyond bosonic string theory: We were hoping to find a string theory with fermionic excitations and a tachyon-free spectrum. In Chapter 8 we constructed a free field theory with bosonic and fermionic fields and supersymmetry on the world-sheet. It came equipped with an N=1 supersymmetric extension of the VIRASORO algebra. The latter is generated by the modes of the VIRASORO field T and its 2-dimensional superpartner G. We shall now try to descend from the field theory to string theory by imposing super-VIRASORO constraints. It shall turn out that the resulting space of stringy wave functions has no negative norm state when the target space is 10-dimensional. Moreover, after imposing an appropriate condition on states, it happens to carry an action of the type IIA/B superalgebras in 10 dimensions. Finally, it is free of tachyonic modes.

10.1 Super-VIRASORO Constraints in the R-Sector

In the second chapter we considered the NAMBU-GOTO action as the fundamental formulation of the bosonic strings, then reformulated it first using POLYAKOV's action, and then from there as a free field theory with VIRASORO constraints. We could follow a similar, logic for the superstring. Not surprisingly, the result would instruct us to use the supermultiplet of the VIRASORO field, i.e., the fields $T, \bar{T}$ along with their superpartners $G, \bar{G}$, as constraints. We shall not go back that far and simply use this strategy now in descending from the free field theory to string theory. Let us begin by discussing the results for the right-moving half of the R-sector. The latter was constructed in chapter 8,

$$\mathcal{H}_{D,\mathrm{R}} = \mathcal{H}_D^{\mathrm{B}} \otimes \mathcal{H}_{D,\mathrm{R}}^{\mathrm{F}}, \tag{10.1}$$

(see eqs. (8.12 and 8.16)) where the construction of the two factors was spelled out in detail. The space $\mathcal{H}_{D,\mathrm{R}}$ carries the action of an N=1 superconformal algebra whose generators L_n, G_n were given in formulas (8.29) and (8.30) above. Before we impose the super-VIRASORO constraints on states in $\mathcal{H}$ let us make two small observations.

As in the case of bosonic strings we want to impose the constraints in a weak sense, i.e., by requiring that physical states are annihilated by all the non-negative modes of the fields T and G. The procedure has some subtlety that was associated with the zero mode L_0. Namely, it turned out that the quantization of the classical constraint l_0 was to be identified with $L_0 - 1$ rather than L_0 itself. The shift originates from the normal ordering ambiguity in L_0. We want to argue now that there is no such ambiguity for the R-sector of our supersymmetric model. In fact, being constructed as a sum of products between the commuting bosonic and fermionic generators a_n^μ and b_n^μ, the operator $G_0 = \sum_n a_{-n}^\mu b_n^\nu \eta_{\mu\nu}$ does not require any normal ordering. Hence, the zero mode G_0 of the current G arises from the quantization of the corresponding classical constraint, i.e., physical states should satisfy $G_0 \,|\,\text{phys.}\rangle = 0$. Moving on to the zero mode of the VIRASORO field we deduce from the commutation relations (8.32) of G that

$$2G_0^2 = \{G_0, G_0\} = 2L_0 - \frac{D}{8} \equiv 2\tilde{L}_0.$$

This suggests that $\tilde{L}_0 = L_0 - D/16$ is the quantum version of the classical constraint l_0 and hence that we should impose $\tilde{L}_0 \,|\,\text{phys.}\rangle = 0$.

These comments along with our previous experience from the bosonic string should suffice to motivate the following construction of the state space in the constrained theory:

$$\mathcal{H}'_{D,R} = \left\{ |\psi\rangle \left| \begin{array}{ll} L_n |\psi\rangle = 0 & n > 0 \\ L_0 |\psi\rangle = \frac{D}{16} |\psi\rangle & \\ G_n |\psi\rangle = 0 & n \geq 0 \end{array} \right. \right\}. \tag{10.2}$$

The only parameter we have not fixed yet is the dimension D of the target space. Once more, we can ask whether there exists a choice of D for which $\mathcal{H}'$ contains no negative norm states. This is the case and requires D to be $D = 10$; see [26, 13, 25]. Since we have seen how such restrictions of the parameters emerge in the case of bosonic string theory, we do not want to go any deeper into these matters here. So, let us accept the choice $D = 10$ and begin studying the wave functions that are obtained from eq. (10.2) after removal of zero norm states. Rewriting the VIRASORO generator L_0 (see eq. (8.20)) as

$$L_0 - \frac{D}{16} = \underbrace{\frac{\alpha'}{4} P^2}_{=-M^2} + \sum_{n=1}^\infty n \underbrace{N_n^B}_{\substack{\text{number of } a\text{'s} \\ \text{with index } -n}} + \sum_{n=1}^\infty n \underbrace{N_n^F}_{\substack{\text{number of } b\text{'s} \\ \text{with index } -n}}, \tag{10.3}$$

we conclude that the mass of physical states must be given by the following simple formula:

$$M^2 = \frac{4}{\alpha'} \sum_{n=1}^{\infty} n N_n,$$

where N_n now counts the number of creation operators b^{μ}_{-n} and a^{μ}_{-n} together. Hence, the spectrum of M^2 is bounded from below by $M^2 \geq 0$, i.e., the R-sector contains no tachyonic modes. Consequently, the lowest lying states are massless, and they must be built out of ground states. We obtain the massless states from

$$|\psi\rangle = |k, \varepsilon_{l=0,\dots,4}\rangle = |k\rangle \otimes \prod_{l=0}^{4} b^{\varepsilon_l}_{0l} |R\rangle. \tag{10.4}$$

The five creation operators b_{0l} were defined in eq. (8.13), and the parameters ε_l can assume the values $\varepsilon_l = 0, 1$. From the constraint above we find

$$0 = G_0 |\psi\rangle = a^{\mu}_0 b^{\nu}_0 \eta_{\mu\nu} |\psi\rangle = \sqrt{\alpha'/2} k_{\mu} b^{\mu}_0 |\psi\rangle \sim (b^0_0 + b^1_0) |\psi\rangle \tag{10.5}$$

in a frame with $k = (1, 1, 0, \dots, 0)$. Note that the operator $b^0_0 + b^1_0 = \sqrt{2} b^{\dagger}_{00}$ is proportional to the annihilation operator $b^{\dagger}_{00}$ we defined in Chapter 8. Hence, states satisfying the physical state condition should not contain the creation operator b_{00}. This leaves us with 32/2=16 states:

$$|k\rangle \otimes \prod_{l=1}^{4} b^{\varepsilon_l}_{0l} |R\rangle. \tag{10.6}$$

Under the action of so(8), these 16 states decompose into two inequivalent irreducible multiplets. Since the generators of so(8) are built from products of two fermionic generators, they respect the natural grading in the R-sector, i.e., they leave invariant the spaces involving an even/odd number of fermionic modes. For the ground states, the associated decomposition has the following form:

1. If $\sum_{l=1}^{4} \varepsilon_l$ is even, the states (10.6) form the spinor representation $\mathbf{8}_S$ of so(8).
2. If $\sum_{i=1}^{4} \varepsilon_l$ is odd, the states (10.6) form the spinor representation $\mathbf{8}_C$ of so(8).

The two representations $\mathbf{8}_S$ and $\mathbf{8}_C$ were discussed briefly in the previous chapter.

10.2 Super-VIRASORO Constraints in the NS-Sector

Let us now repeat the analysis of string theory states for a chiral theory in the NS-sector. The underlying field theory state space is given by

$$\mathcal{H}_{D,NS} = \mathcal{H}^B_D \otimes \mathcal{H}^F_{D,NS}. \tag{10.7}$$

It carries the action of an N=1 superconformal algebra that is generated by L_n and the modes G_r with $r \in \mathbb{Z} + \frac{1}{2}$. The space of physical states is constructed by imposing the following physical state conditions:

$$
\mathcal{H}'_{D,\tilde{a},\,\mathrm{NS}} = \left\{ |\psi\rangle \,\middle|\,
\begin{array}{ll}
L_n |\psi\rangle = 0 & n > 0 \\
L_0 |\psi\rangle = \tilde{a} |\psi\rangle & \\
G_r |\psi\rangle = 0 & r \geq \frac{1}{2}
\end{array}
\right\}.
\tag{10.8}
$$

We point out that the constant $\tilde{a}$ cannot be fixed by arguments similar to the ones we discussed for the R-sector. In this respect, the analysis in the NS-sector is closer to our construction of the bosonic string. As in the purely bosonic theory, the value of $\tilde{a}$ is fixed along with the value of the dimension D by a no ghost theorem [26, 13, 25]. It asserts that the space $\mathcal{H}'_{D,\tilde{a},\mathrm{NS}}$ contains no states of negative norm provided that $D = 10$ and $\tilde{a} = \frac{1}{2}$. Fortunately, the critical dimension turns out to be the same as in the R-sector. The space of physical states is obtained from $\mathcal{H}'_{10,1/2,\mathrm{NS}}$ by removing all zero norm states. From the expression for the mass

$$
M^2 = -\frac{2}{\alpha'} + \frac{4}{\alpha'} \sum_{n=1}^{\infty} \frac{n}{2} \underbrace{N_{\frac{n}{2}}}_{\substack{\text{number of modes} \\ \text{with index } -n/2}}
\tag{10.9}
$$

we conclude that the NS-sector spectrum contains a tachyonic mode of the form

$$
|\psi\rangle = |k\rangle \otimes |0\rangle \quad \text{where} \quad k^2 = \frac{2}{\alpha'}
$$

and $|0\rangle$ is the NS ground state. This is certainly disappointing. Before we address the issue, let us briefly discuss massless states. These are obtained from $|\psi\rangle = \xi_\mu b^\mu_{-\frac{1}{2}} |k\rangle \otimes |0\rangle$, such that

$$
G_{\frac{1}{2}} |\psi\rangle = \eta_{\mu\nu} a_0^\mu b^\nu_{\frac{1}{2}} |\psi\rangle = \xi_\mu k^\mu |\psi\rangle \overset{!}{=} 0,
\tag{10.10}
$$

$$
\langle \psi | \psi \rangle = \xi^2 = 0 \text{ if } \xi_\mu \sim \lambda k_\mu.
\tag{10.11}
$$

Hence, the polarization is transverse $k \cdot \xi = 0$. Once we remove the longitudinally polarized state of zero norm, we are left with the content of a massless vector multiplet $\mathbf{8}_V$ for so(8).

10.3 GSO Projection and Spectrum of Superstrings

Things have not quite turned out the way we had hoped. The chiral spectrum we discussed in the previous two sections has two major problems. To begin, the NS-sector does contain a tachyonic mode. Furthermore, there is no space-time supersymmetry between massless space-time bosons in the representation $\mathbf{8}_V$ and fermions in the representation $\mathbf{8}_S \oplus \mathbf{8}_C$. Let us point out that the unwanted tachyonic mode contains no fermionic creation operators. On the other hand, the desired massless vector multiplet in the NS-sector is built from states involving one fermionic mode. In the R-sector, one of the two 8-dimensional multiplets contains an odd number of fermionic creators, while the other includes states with an even number of creation operators. These observations may suggest the following way out: What if we simply project on states that contain an odd number of fermionic generators? That would remove the tachyonic mode along with one of the fermionic octoplets. As bizarre as the prescription might seem at first, it passes stringent consistency conditions and became known as the GLIOZZI-SCHERK-OLIVE (GSO) projection [24]. In the next few paragraphs we will implement this idea more concretely. A few comments on possible consistency checks are added at the end of the chapter.

In order to project on odd states we shall introduce two operators $\Gamma_\pm$ that commute with all bosonic generators a_n^μ and anti-commute with the fermionic ones, i.e., $\Gamma_\pm b_l^\mu = -b_l^\mu \Gamma_\pm$ where $l = 0, \pm 1/2, \pm 1, \ldots$. On ground states, the operators $\Gamma_\pm$ are defined by

$$\Gamma_\pm \,|\, k\rangle \otimes |\, 0\rangle \;=\; -\,|\, k\rangle \otimes |\, 0\rangle\,, \qquad \Gamma_\pm \,|\, k\rangle \otimes |\, \mathrm{R}\rangle = \pm\,|\, k\rangle \otimes |\, \mathrm{R}\rangle\,. \qquad (10.12)$$

This defines $\Gamma_\pm$ uniquely. Although it is needed only in later chapters, we would like to spell out an explicit construction of these operators in terms of the fermionic generators. There are many ways to accomplish this, but here we shall adopt the following somewhat complicated looking solution:[1]

$$\Gamma_\pm^{\mathrm{NS}} \;=\; -\,(-1)^{\,\mathrm{i}\sum_{j=0}^4 \sum_{r=1/2}^\infty \left(b_{-r}^{2j}b_r^{2j+1} - b_{-r}^{2j+1}b_r^{2j}\right)}{}_{b_r^0 \to -\mathrm{i} b_r^0}\,, \qquad (10.13)$$

$$\Gamma_\pm^{\mathrm{R}} \;=\; \pm 32\, b_0^0 b_0^1 \cdots b_0^9\, (-1)^{\,\mathrm{i}\sum_{j=0}^4 \sum_{n=1}^\infty \left(b_{-n}^{2j}b_n^{2j+1} - b_{-n}^{2j+1}b_n^{2j}\right)}{}_{b_n^0 \to -\mathrm{i} b_n^0}\,. \qquad (10.14)$$

The subscript on the exponents instructs us to rotate all fermionic modes that are associated with the time-like direction by a factor $-\mathrm{i}$. We also note that the product of zero modes in the second line mimics the construction of γ^{11}. The factor 32 originates from our normalization of the zero modes, which differs from the normalization of γ matrices. We leave it as an exercise to check that the operators (10.13) and (10.14) indeed satisfy all the properties we demanded.

Even from our initial characterization of $\Gamma_\pm$ it is obvious that these operators descend to the states in $\mathcal{H}'$. The GSO projection amounts to keeping only those states that have eigenvalue $+1$ under the action of $\Gamma_\pm$. Thereby, we remove the

---

[1] These explicit expressions will feature briefly in Chapter 11 and more profoundly at the end of Chapter 17. For the rest of the book the simple characterization of $\Gamma_\pm$ we gave before suffices.

tachyonic states and half of the ground states in the R-sector. Before we go into more detail, let us now combine the left- and right-moving sectors of the theory. Before GSO projection the state space is then given by

$$\mathcal{H}' = \mathcal{H}'_{\text{NSNS}} \oplus \mathcal{H}'_{\text{RR}} \oplus \mathcal{H}'_{\text{RNS}} \oplus \mathcal{H}'_{\text{NSR}}. \tag{10.15}$$

It admits the action of two types of projection operators:

$$\Pi_A = \Pi_+ \bar{\Pi}_- \quad \text{or} \quad \Pi_A = \Pi_- \bar{\Pi}_+, \tag{10.16}$$

$$\Pi_B = \Pi_+ \bar{\Pi}_+ \quad \text{or} \quad \Pi_B = \Pi_- \bar{\Pi}_-, \tag{10.17}$$

where $\Pi_\pm = (1 + \Gamma_\pm)/2$, etc. Physical states that survive the projection with the operator Π_A constitute the spectrum of type IIA superstring theory. Those that remain after acting with Π_B, on the other hand, form the space of wave functions for the type IIB superstring.

We begin our more detailed investigation of the massless states with the type IIB theory. After applying the GSO projection to the NSNS-sector, the tachyon is removed and the massless states become

$$8_V \times 8_V = \underbrace{\mathbf{1}}_{\text{dilaton}} \oplus \underbrace{\mathbf{28}}_{\text{Kalb-Ramond}} \oplus \underbrace{\mathbf{35}_V}_{\text{graviton}}. \tag{10.18}$$

On the right-hand side, we display the usual decomposition of the product of vectors into the trace, antisymmetric, and symmetric traceless tensors. The individual parts correspond to the dilaton Φ, the KALB-RAMOND field $B^{\alpha\beta}$, and the metric or graviton $G^{\alpha\beta}$, respectively. In the RR-sector, the massless multiplets become

$$8_S \times 8_S = \underbrace{\mathbf{1}}_{0-\text{form}} \oplus \underbrace{\mathbf{28}}_{2-\text{form}} \oplus \underbrace{\mathbf{35}_V}_{\text{self-dual 4-form}}, \tag{10.19}$$

which correspond to 0, 2-, and 4-form RR-fields, respectively. Relevant background on the decomposition of the tensor product can be found, e.g., in [63]. The NSNS- and RR-sectors represent space-time bosons, whereas the RNS- and NSR-sectors are space-time fermions. We first give the massless multiplets of the RNS-sector,

$$8_V \times 8_S = \underbrace{\mathbf{8}_S}_{\text{dilatino}} \oplus \underbrace{\mathbf{56}_S}_{\text{gravitino}}, \tag{10.20}$$

which are the left-handed dilatino and gravitino. In the NSR-sector the massless multiplets are

$$8_S \times 8_V = \underbrace{\mathbf{8}_S}_{\text{dilatino}} \oplus \underbrace{\mathbf{56}_S}_{\text{gravitino}}, \tag{10.21}$$

which are also left-handed dilatino and gravitino multiplets. Observe that the spectrum is chiral so that the massless multiplets of N=2 supersymmetry form the spectrum of type IIB supergravity. Note that there are as many bosons as fermions with $m = 0$.

We now turn our attention to the spectrum of the type IIA superstring, which comes from the $|\psi\rangle \in \mathcal{H}'$ imposing $\Pi_A |\psi\rangle = |\psi\rangle$. In the NSNS-sector, again, the tachyon is removed, and the massless spectrum is as in type IIB before. The massless multiplets in the RR-sector become

$$\mathbf{8}_C \times \mathbf{8}_S = \underbrace{\mathbf{8}_V}_{\text{1-form}} \oplus \underbrace{\mathbf{56}_V}_{*\text{3-form}} , \tag{10.22}$$

which are the 1- and 3-form RR-fields, respectively. In the RNS-sector we have

$$\mathbf{8}_V \times \mathbf{8}_S = \underbrace{\mathbf{8}_S}_{\text{dilatino}} \oplus \underbrace{\mathbf{56}_S}_{\text{gravitino}} , \tag{10.23}$$

which are the left-handed dilatino and gravitino. The NSR-sector now has multiplets with different chirality,

$$\mathbf{8}_C \times \mathbf{8}_V = \underbrace{\mathbf{8}_C}_{\text{dilatino}} \oplus \underbrace{\mathbf{56}_C}_{\text{gravitino}} , \tag{10.24}$$

i.e., they are right-handed rather than left-handed. In other words, the spectrum of type IIA theory is non-chiral and N=2 supersymmetric.

We have verified that the massless states of both type IIA and type IIB superstring theory form multiplets of the type IIA and type IIB POINCARÉ superalgebras. It is possible to show that the same is true for all the massive excitations in the spectrum. This target space supersymmetry was not manifest in our construction of the spectrum, and it needed some work to collect all the bosonic multiplets into representations of the POINCARÉ superalgebra. This is a bit similar to what we saw in the light-cone quantization of the bosonic string. In that approach to quantizing the string, the target space LORENTZ symmetry was not manifest, and it had to be established by setting $D = 26$. The approach to quantizing the superstring we outlined above is known as NSR formalism. It is manifestly supersymmetric on the world-sheet, but the target space supersymmetry emerges only at the very end of the construction. There is another approach to superstrings, known as the GREEN-SCHWARZ formalism, in which target space (but not world-sheet) supersymmetry is manifest. This formalism, however, cannot be quantized without breaking at least part of the target space supersymmetry.

Let us conclude this chapter with a few more comments on the GSO projection. It may seem like some kind of a dirty trick, but it really turns out to be very strongly constrained by the consistency of the resulting theory. Let us suppose for a moment that we had tried to remove the tachyon of bosonic string theory with a similar idea, namely by projecting out all states that involve an even number of bosonic oscillators. This would then have left us with the massless modes of the bosonic strings at the bottom of the spectrum. In such a "theory" we could have computed, e.g., a 4-point scattering amplitude between massless bosonic strings. We have done a similar computation for four tachyon modes of the closed string before. Performing the same kind of calculation for four massless modes we obtain an amplitude that contains once more several Γ functions, and we may then look for the poles of the latter in the MANDELSTAM variable t. The results are essentially the same as before: the first pole appears at $t = M^2 = -4/\alpha'$, i.e., arises from a tachyon. In other words, removing the tachyon by hand from the spectrum of the closed bosonic string would not have helped at all. It comes back into the theory as one of the modes that is exchanged in string scattering experiments. In this sense, projecting the bosonic string onto a tachyon-free spectrum is inconsistent. This is very different from what happens for the superstring. A similar computation involving the massless modes of the type IIA/B superstring spectrum results in an amplitude that has no tachyonic pole in the t-channel. Hence, the spectrum we obtained after GSO projection is consistent; i.e., scattering experiments of states within the spectrum never produce any new states that were not included.

10.4 Type IIA/B Supergravity Theories

Having obtained a tachyon-free supersymmetric spectrum of the type II/B superstring theory we can now move ahead and compute the low-energy effective action for massless fields. We have performed similar computations for massless modes of the open bosonic string in Chapter 7. Clearly, the analogous computations for the closed bosonic superstring are a bit more involved, but in principle they follow the same scheme, in particular in the NS-sector. We will not perform the relevant calculations of scattering amplitudes here, but simply state the results. To leading order in the parameter α', the amplitudes of the massless modes of the closed type IIA/B superstring may be encoded in the action of 10-dimensional type IIA/B supergravity. There are a number of terms in these actions that involve only NS-sector fields. These are identical for the two 10-dimensional N=2 supergravities and read

$$S_{NS} = \frac{1}{2\kappa_{10}^2} \int d^{10}x \sqrt{-g}\, e^{-2\phi} \left(R(g) + 4\partial_\mu \phi \partial^\mu \phi - \frac{1}{2}|H_3|^2 \right). \qquad (10.25)$$

Here, the metric and dilaton are denoted by g and ϕ, respectively. The B-field enters through the NSNS 3-form flux $H_3 = dB$, and $R(g)$ denotes the usual scalar curvature. The factor $\exp(-2\phi)$ is characteristic for closed strings as we shall explain later.

All remaining bosonic terms of type IIA/B supergravity contain RR-fields. Since the two theories possess a different set of RR-fields, the actions also look slightly different. Let us begin by spelling out the remaining terms for the type IIA theory,

$$S_R^{IIA} = -\frac{1}{4\kappa_{10}^2} \int d^{10}x \sqrt{-g} \left(|F_2|^2 + |\tilde{F}_4|^2 + B_2 \wedge F_4 \wedge F_4 \right), \quad (10.26)$$

where

$$\tilde{F}_4 = dC_3 - C_1 \wedge H_3 = F_4 - C_1 \wedge H_3 \quad (10.27)$$

and $F_{p+2} = dC_{p+1}$ denote the field strength of the RR potentials C_{p+1} as before. Recall that p is even in type IIA theory. The final term in the action (10.26) is often referred to as the CHERN-SIMONS term. More generally, the CHERN-SIMONS terms are contributions to the action of gauge fields that cannot be written in terms of the field strength F alone but rather involve gauge potentials C. Of course, such terms must still be gauge invariant. In the case of the last term of eq. (10.26), a gauge transformation $\delta B_2 = d\Lambda_1$ indeed acts trivially:

$$\int d^{10}x \, d\Lambda_1 \wedge F_4 \wedge F_4 = - \int d^{10}x \, \Lambda_1 \wedge d(F_4 \wedge F_4) = 0.$$

Here we have used STOKES's theorem assuming that we can neglect boundary terms at infinity, the LEIBNIZ rule for total derivatives, and finally the BIANCHI identity $dF_4 = 0$. Gauge invariance of the integral involving $|\tilde{F}_4|^2$ is a little more tricky. Under a gauge transformation, the potential C_1 behaves as $\delta C_1 = d\Lambda_0$ so that one would expect $\delta \tilde{F}_4 = -d\Lambda_0 \wedge H_3 \neq 0$. In order to save the gauge invariance of $\tilde{F}_4$ one needs to make these gauge transformations act non-trivially on the gauge potential C_3 of the 4-form field strength F_4 as

$$\delta C_3 = \Lambda_0 \wedge H_3 \quad \text{such that} \quad \delta\tilde{F}_4 = d(\Lambda_0 \wedge H_3) - d\Lambda_0 \wedge H_3 = 0.$$

Note that the terms involving gauge potentials disappear from the action we if set the B-field to be trivial, i.e., $B = 0$.

In the case of type IIB supergravity, the corresponding part of the action reads

$$S_R^{IIB} = -\frac{1}{4\kappa_{10}^2} \int d^{10}x \sqrt{-g} \left(|F_1|^2 + |\tilde{F}_3|^2 + \frac{1}{2}|\tilde{F}_5|^2 + C_4 \wedge H_3 \wedge F_3 \right), \quad (10.28)$$

where

$$\tilde{F}_3 = F_3 - C_0 \wedge H_3 \quad , \quad \tilde{F}_5 = F_5 - \frac{1}{2}C_2 \wedge H_3 + \frac{1}{2}B_2 \wedge F_3. \quad (10.29)$$

All the comments we made above on the definition of the various terms and on gauge invariance apply to the type IIB action as well. Based on what we know about superstrings and in particular about the computation of scattering amplitudes involving RR-fields it is not really possible to explain all the features of the above actions easily. But let us add at least one comment that may help to appreciate the fact that all terms in the above action contain an even number of RR-fields. This has a simple explanation. The vertex operators for RR-fields contain the RR ground state, which creates a square branch point on the world-sheet; see exercise 21. Hence, non-zero correlation functions must contain an even number of RR vertex operators. A more detailed discussion of type IIA/B supergravities and their relation to superstring theory can be found in chapter 12 of [56].

Exercises

Problem 27. *Verify that the operators $\Gamma_{\pm}$ that are defined through eqs. (10.13) and (10.14) anticommute with all fermionic creation operators and satisfy eqs. (10.12).*

11

Branes in Type II Superstring Theory

Having constructed our first examples of tachyon-free closed string models we now turn to the description of branes and open strings in type II superstring theory. We shall begin with a brief discussion of solitonic branes in type II supergravity. This part is rather descriptive. In particular, we do not intend to explain how these solutions are found. We shall then turn to the string theoretic analogue of solitonic branes. As we explained in the introductory chapter branes in string theory are objects on which open strings can end. We shall use this as a definition of D-branes in string theory and verify that the resulting objects share many features with the solitonic branes in supergravity. Finally, gauge theories on the world-volume of D-branes are discussed briefly.

Our quantization of open strings in this section will proceed through the light-cone gauge rather than the covariant formalism. This will give us a good opportunity to review the basic ideas of light-cone gauge quantization. But there is another reason for us to do so. In the last chapters we have spent some time to verify that the massless states of type II string theories form full multiplets of the POINCARÉ superalgebra. Extending this analysis to massive sectors would be a rather difficult task. Here we want to verify that all eigenspaces of the mass operator for open strings in the type II theory possess the same number of bosonic and fermionic components. To reach that goal, the light-cone gauge has clear advantages since it allows for a direct and simple construction of physical states.

11.1 Solitonic Branes in Type II Supergravity

At the end of the previous section we have seen N=2 supersymmetric supergravities in 10 dimensions. As we argued there, these theories may be considered as higher dimensional generalizations of EINSTEIN-MAXWELL theory. In fact, the action of type IIA/B supergravities contains a number of higher form potentials C_{p+1}. From the study of 4-dimensional gravity it is well known that there exist special solutions that describe charged black holes. These so-called REISNER-NORDSTRÖM solutions for the metric are labeled by two parameters m and q,

$$ds^2 = -dx_0^2 \left(1 - \frac{2m}{\rho} + \frac{q^2}{\rho^2}\right) + d\rho^2 \left(1 - \frac{2m}{\rho} + \frac{q^2}{\rho^2}\right)^{-1} + \rho^2 d\Omega^2, \quad (11.1)$$

where $d\Omega^2 = d\theta^2 + \sin^2 d\phi^2$ is the line element on the 2-dimensional unit sphere. This metric admits two KILLING horizons at the zeroes of the factor $(1 - 2m/\rho + q^2/\rho^2)$. When $m < |q|$, the two horizons disappear, and we have a naked singularity that is forbidden by cosmic censorship. The case $m = |q|$ where the two horizons come together is known as an extremal REISNER-NORDSTRÖM black hole. In this case, the metric can be brought into the simpler form

$$ds^2 = -H(r)^{-2}dx_0^2 + H(r)^2(dr^2 + r^2 d\Omega^2) \quad \text{where} \quad H(r) = \left(1 + \frac{m}{r}\right), \quad (11.2)$$

and the variable r is related to ρ through $\rho = r + m$. We are now going to discuss solutions of a similar form for 10-dimensional type IIA/B supergravity.

Let us recall that type II supergravity contains certain higher form fields C_{p+1} in addition to the usual dilaton ϕ, the KALB-RAMOND 2-form potential B, and the metric. We remind readers that the action for the bosonic part of the 10-dimensional supergravity theory is[1]

$$S = \frac{1}{G_{10}} \int d^{10}x \sqrt{-g} \left(R - \frac{1}{2}(\partial\phi)^2 - \sum_p \frac{1}{2(p+2)!} e^{\frac{p-3}{2}\phi} \left(dC_{p+1}\right)^2\right). \quad (11.3)$$

In the above, we have dropped the B field and all terms involving fermionic fields since these are not relevant for the solutions we are going to describe. In order to spell out static solutions of the action S we pick some integer $-1 \le p \le 6$, and we split the 10 coordinates into the two pieces $x = (x_0, \ldots, x_p)$ and $y = (x_{p+1}, \ldots, x_9)$. The supergravity action S is extremized by

$$ds^2 = f_p^{\frac{p-7}{8}} dx^2 + f_p^{\frac{p+1}{8}} \left(dr^2 + r^2 d\Omega_{8-p}^2\right),$$

$$e^{2\phi} = f_p^{\frac{3-p}{2}}, \quad (11.4)$$

$$C_{p+1} = f_p^{-1} dx_0 \wedge \cdots \wedge dx_p.$$

---

[1] The action is displayed in the so-called EINSTEIN frame. In order to relate it to eqs. (10.25), (10.26), and (10.28) in the previous chapter one has to redefine the metric g through $g \exp(-\phi/2)$.

Here, $\Omega_{d-1}^2 = 2\pi^{\frac{d}{2}}\Gamma(\frac{d}{2})$ is the volume of the $(d-1)$-dimensional unit sphere, $d\Omega_{d-1}^2$ is its metric, and we have set $r = |y|$. The quantity f_p, finally, is defined by

$$f_p := 1 + \frac{G_{10}T_p}{(7-p)\Omega_{8-p}} \frac{1}{r^{7-p}}. \tag{11.5}$$

Note that f_p contains one free parameter T_p. The solitonic solution (11.4) describes an object with mass density T_p and some C_{p+1}-charge density Q_{p+1} that is localized along the $(p+1)$-dimensional hyperplane given by the equation $y = 0$. In order to understand the power law decay of these solutions, we expand the last equation of eqs. (11.4) for large radial distance r:

$$C_{p+1} \sim -\frac{G_{10}T_p}{(7-p)\Omega_{8-p}} \frac{1}{r^{7-p}} dx_0 \wedge \cdots \wedge dx_p.$$

Here we have dropped a constant term since it is pure gauge, i.e., it does not contribute to the field strength dC_{p+1}. We can now easily recognize a COULOMB-type law. Recall that the COULOMB potential of a point-like charge in a n-dimensional space behaves as $V(r) \sim r^{2-n}$ for $n \geq 3$. The same law holds for a p-dimensional charged object in a $(p+n)$-dimensional space. In our case $p + n = 9$ so that the COULOMB law takes the form $V(r) = Q_p/r^{7-p}$, in agreement with the behavior of the $p+1$ form potential C_{p+1}. We can also read off that the charge density Q_{p+1} of our solution is related to the mass density T_p through

$$Q_{p+1} = \frac{G_{10}T_p}{(7-p)\Omega_{8-p}}. \tag{11.6}$$

This relation between charge and mass density has a remarkable consequence. Consider a configuration of two identical and parallel branes that are placed at a certain distance from each other. The gravitational force attracts these two branes to each other. On the other hand, having equal (rather than opposite) charges, the exchange of C_{p+1} gauge bosons makes the branes want to separate. If charge and mass density are related through equation (11.6), then the two branes turn out not to exert any net force onto each other, i.e., their gravitational attraction is balanced exactly by the COULOMB repulsion.

Let us recall that the forms C_{p+1} with p odd appear in type IIB theory, while those with p even are fields in IIA supergravity. Hence, the solutions (11.4) with $p = -1, 1, 3, 5$ are objects of IIB supergravity, while those with $p = 0, 2, 4, 6$ represent objects in the IIA theory. To be precise, we stress that all the fermionic fields, the B field and the form fields $C_{p'+1}$ with $p' \neq p$, may consistently be set to zero. Solutions of this type are known as solitonic p-branes.

There exists a second set of solutions of a very similar form that describe magnetically charged objects. In these solutions mass and magnetic charge density are localized along a $(7-p)$-dimensional hyper-manifold in the 10-dimensional space time. The corresponding $q+1 = 7-p$-form potential $\tilde{C}_{q+1}$ gives rise to a $q+2$ form field strength $d\tilde{C}_{q+1}$. Its HODGE dual $*d\tilde{C}_{q+1}$ is a $8-q$-form with $7-q$-form potential $C_{7-q} = C_{p+1}$. These solutions are similar to the the DIRAC monopole solutions of MAXWELL theory in $3+1$ dimensions. In that case, the gauge field is a 1-form potential A_1, and it possesses a 2-form field strength F_2 whose HODGE dual $*F_2$ is a 2-form as well so that it has a 1-form potential, the famous DIRAC monopole. Let us recall that in quantum theory the existence of a monopole implies charge quantization. The same is true for the higher dimensional analogues of the DIRAC monopole.

The only form field that is not sourced by any of these p-brane solutions is the KALB-RAMOND field B. One may wonder whether there are other solutions of type IIA/B supergravity in which B is non-zero and all the RR-forms C_{p+1} vanish. This is indeed the case. In fact, there exists two types of solutions that come with the KALB-RAMOND field. One of them describes an object stretching out along a $(1+1)$-dimensional hyperplane. It is known as the fundamental string solution. The dual solution belongs to a $(5 + 1)$-dimensional hyperplane. We will come across these so-called NS5-branes later. For the moment, however, they are not relevant for us. NS5-branes can be modeled in string theory [40], but the way this is done has very little in common with how one implements the classical p-brane solution (11.4).

11.2 Open Strings in Type II Superstring Theory

The following section contains a brief construction of the space of physical states for open superstrings. As in our discussion of open bosonic strings in Chapter 6, we shall impose NEUMANN boundary conditions along the first $p + 1$ directions. DIRICHLET boundary conditions are imposed for the remaining $9 - p$ coordinate fields.

The Bosonic Sector

For the bosonic fields in our type II superstring theory, the boundary conditions are the same as in the bosonic string. It is useful, however, to rewrite those conditions in terms of the chiral fields $J_\pm^\mu(\sigma^\pm) = i\partial_\pm X^\mu$. This will help us later to find the appropriate boundary conditions for fermions. Let us recall that we impose NEUMANN boundary conditions for the first $p + 1$ coordinates and DIRICHLET boundary conditions for the remaining ones. The conditions read

$$\partial_\sigma X^\mu(\tau,\sigma)\big|_{\sigma=0,\pi} = 0 \quad \leftrightarrow \quad J_+^\mu(\sigma^+) = \left. J_-^\mu(\sigma^-)\right|_{\sigma=0,\pi} \text{ for } \mu = 0,\dots,p,$$

$$(11.7)$$

$$\partial_\tau X^\mu(\tau,\sigma)\big|_{\sigma=0,\pi} = 0 \quad \leftrightarrow \quad J_+^\mu(\sigma^+) = \left. -J_-^\mu(\sigma^-)\right|_{\sigma=0,\pi} \text{ for } \mu = p+1,\dots,9.$$

$$(11.8)$$

In passing from the familiar left-hand side to our new formulation of boundary conditions we simply had to insert $\partial_\sigma = \partial_+ - \partial_-$ and $\partial_\tau = \partial_+ + \partial_-$. Note that DIRICHLET and NEUMANN boundary conditions differ by a sign that describes how the left- and right-moving fields J_+ and J_- are "glued" along the boundary. Bosonic fields solving these boundary conditions must be of the form

$$X^\mu(\sigma^-, \sigma^+) = x^\mu - i\alpha' p^\mu \tau - i\sqrt{\frac{\alpha'}{2}} \sum_{n \neq 0} \left(\frac{a_n^\mu}{n} e^{-in\sigma^-} + \frac{\bar{a}_n^\mu}{n} e^{-in\sigma^+} \right) \quad (11.9)$$

with

$$a_n^\mu = \bar{a}_n^\mu =: \Omega_p(\bar{a}_n^\mu) \quad \text{for} \quad \mu = 0, \ldots, p, \quad (11.10)$$

$$a_n^\mu = -\bar{a}_n^\mu =: \Omega_p(\bar{a}_n^\mu) \quad \text{for} \quad \mu = p+1, \ldots, 9 \quad (11.11)$$

and $p^\mu = x^\mu = 0$. On the right-hand side we have defined an automorphism Ω_p that acts on the left-moving modes $\bar{a}_n^\mu$. This automorphism describes the non-trivial identification between left- and right-moving modes. It depends on the dimension p of the branes, i.e., on the choice of boundary conditions.

The Fermionic Sector

Let us now turn to the world-sheet fermions $\psi_\pm^\mu$. In order to preserve the symmetry between bosonic and fermionic fields we impose the following boundary conditions:

$$\psi_+^\mu = \psi_-^\mu\big|_{\sigma=0} \quad \psi_+^\mu = \epsilon\psi_-^\mu\big|_{\sigma=\pi} \quad \text{for } \mu = 0, \ldots, p, \quad (11.12)$$

$$\psi_+^\mu = -\psi_-^\mu\big|_{\sigma=0} \quad \psi_+^\mu = -\epsilon\psi_-^\mu\big|_{\sigma=\pi} \quad \text{for } \mu = p+1, \ldots, 9. \quad (11.13)$$

As in the case of closed strings we have left the freedom to choose $\epsilon = \pm$. The case $\epsilon = 1$ corresponds to the RAMOND sector, whereas $\epsilon = -1$ belongs to the NEVEU-SCHWARZ sector. The general solution of the equation of motion with the above boundary conditions takes the form

$$\psi_\pm^\mu(\sigma^\pm) = \begin{cases} \sqrt{i} \sum_n b_{n,\pm}^\mu e^{-in\sigma^\pm} \\ \sqrt{i} \sum_r b_{r,\pm}^\mu e^{-ir\sigma^\pm} \end{cases}, \quad (11.14)$$

where we have used the notation $b_{n,+}^\mu = b_n^\mu$ and $b_{n,-}^\mu = \bar{b}_n^\mu$ for convenience. The boundary conditions imply that the modes fulfill

$$b_n^\mu = \bar{b}_n^\mu =: \Omega_p(\bar{b}_n^\mu) \quad \text{for} \quad \mu = 0, \ldots, p, \quad (11.15)$$

$$b_n^\mu = -\bar{b}_n^\mu =: \Omega_p(\bar{b}_n^\mu) \quad \text{for} \quad \mu = p+1, \ldots, 9 \quad (11.16)$$

and similarly for the modes b_r^μ in the NS-sector. Once more we have described the boundary conditions through an automorphism Ω_p of the algebra that is generated by the left-moving modes $\bar{b}_n^\mu$.

The Light-Cone Gauge

These relations have the consequence that the bosonic terms of the stress energy tensor

$$T_{\pm\pm} = \frac{1}{\alpha'} J_\pm^\mu J_{\pm,\mu} - \frac{1}{2} \psi_\pm^\mu \partial_\pm \psi_{\pm,\mu} \tag{11.17}$$

satisfies the following gluing condition along the boundary of the upper half plane:

$$T_{++} = T_{--} \qquad \text{for } \sigma = 0, \pi.$$

The relation implies that the VIRASORO generators L_n and $\bar{L}_n$ coincide, i.e., that there is a single set of VIRASORO generators in our boundary theory. Therefore, we can implement the VIRASORO constraints by setting $a_n^+ = 0$ for all $n \neq 0$ and $x^+ = 0$. With this gauge choice the solution X^+ reads

$$X^+(\tau, \sigma) = p^+ \tau. \tag{11.18}$$

In contrast to the theory of closed strings, the chosen gauge condition fixes all gauge freedom. Hence, there is no analogue of the level matching condition. Putting all this together we conclude that the bosonic state space is generated by a_n^α for $\alpha = 2, \ldots, 9$ and $n < 0$ out of the ground states $|k^2, \ldots, k^p, p^+\rangle$.

Our choice of boundary conditions for the fermion fields also ensures that the superpartner

$$G_\pm = i \sqrt{\frac{2}{\alpha'}} \psi_\pm^\mu \partial_\pm X_\mu = \sqrt{\frac{2}{\alpha'}} \psi_\pm^\mu J_{\pm,\mu} \tag{11.19}$$

of the stress energy tensor satisfies the following conditions at the boundaries:

$$G_+ = G_- \big|_{\sigma=0}, \quad G_+ = \epsilon G_- \big|_{\sigma=\pi}. \tag{11.20}$$

Once again it implies that the modes of G_+ and G_- coincide up to a sign. We can now use the super-diffeomorphism invariance of the theory, i.e., the constraint that $G_+ \sim G_-$ vanishes, in order to implement the vanishing of the light-cone modes $b_n^+ = \bar{b}_n^+ = 0, b_r^+ = \bar{b}_r^+ = 0$, which then sets

$$\psi^+(\tau, \sigma) = 0. \tag{11.21}$$

The fermionic state space is generated by the remaining modes $b_{n,r}^\alpha$ with $n, r < 0$, and $\alpha = 2, \ldots, 9$ from the ground state $|NS\rangle$ of the NS-sector and the ground states of the R-sector:

$$\prod_{l=0}^{4} b_{0l}^{\varepsilon_l} |R\rangle \qquad \text{with} \qquad \varepsilon_l \in \{0, 1\}\,.$$

Note that we did not include $l = 0$ into the product of zero modes since our gauge $b_0^+ = 0$ implies that $b_{00} = b_0^- \sim p_\alpha b_0^\alpha / p^+$ can be expressed in terms of zero modes $b^\alpha, \alpha = 2, \ldots, 9$ and hence does not create new states.

In order to complete the construction of the space of physical states of open superstrings it remains to impose the GSO projection. Following eqs. (10.13) and (10.14), the projection operators $\bar\Pi_\pm = (1 + \bar\Gamma_\pm)/2$ can be written out in terms of fermionic modes. These explicit formulas can now be used to verify that

$$\bar\Pi_\pm^R = \Omega_p(\bar\Pi_\pm^R) \quad \text{for} \quad p \text{ odd}, \tag{11.22}$$

$$\bar\Pi_\pm^R = \Omega_p(\bar\Pi_\mp^R) \quad \text{for} \quad p \text{ even}. \tag{11.23}$$

Here, Ω_p is the gluing morphism we have introduced above. It is not difficult to see that the automorphism Ω_p does not change the parity of the exponent of (-1) in the definition of $\bar\Gamma_\pm$. In particular, it therefore leaves the projection operator in the NS-sector invariant. In the Ramond sector, the automorphism multiplies $9 - p$ of the zero modes $\bar b_0^\mu$ by -1. Consequently,[2]

$$\Omega_p(\bar b_0^2 \bar b_0^3 \cdots \bar b_0^9) = (-1)^{9-p} \bar b_0^2 \bar b_0^3 \cdots \bar b_0^9.$$

This shows that Ω_p indeed switches between $\bar\Pi_+$ and $\bar\Pi_-$ if p is even. Recall that the type IIA theory employs opposite projectors $\Pi_\pm$ and $\bar\Pi_\mp$; see eq. (10.16). Hence, the type A projection is consistent with the gluing morphisms Ω_p when p is even, i.e.,

$$\Pi_A = \Pi_+\Omega_p\left(\bar\Pi_-\right) = \Pi_+\bar\Pi_+ = \Pi_+\Pi_+ = \Pi_+$$

for p even. Here we took the type A projection for closed strings from eq. (10.16), and applied the gluing automorphism Ω_p and the identifications (11.15) and (11.16) between left- and right-moving fermionic modes. The last equality is the projection property of Π_+. Had we performed this computation for odd p instead, we would have obtained $\Pi_A = \Pi_+\Pi_- = 0$. Similarly, a consistent GSO-projected open string theory for type IIB requires p to be odd. Thereby, we have recovered our geometric result that Dp-branes with $p = 0, 2, 4, \ldots$ exist in the type IIA theory, while those with $p = -1, 1, 3, \ldots$ are restricted to type IIB theory.

---

[2] In our light-cone gauge $\bar b_0^0 \bar b_0^1 = -\bar b_0^0 \bar b_0^0 = 1/2$ so that we conclude $32\bar b_0^0 \bar b_0^0 \bar b_0^1 \bar b_0^2 \cdots \bar b_0^9 = 16 \bar b_0^2 \cdots \bar b_0^9$.

11.3 Supersymmetry of the Spectrum

The theory we are investigating is claimed to be supersymmetric. This implies in particular that all the eigenspaces of the mass operator should contain equal numbers of bosonic and fermionic modes. Since space-time bosons arise from the NS-sector and space-time fermions are associated with the R-sector, we will have to count states in the two sectors separately so that we can compare the number of states level by level.

This counting needs a bit of preparation. To begin, let us consider a single bosonic direction, i.e., a set of oscillators $a_n, n \in \mathbb{Z}$, with the usual commutation relations. We would like to count the states that they create from the ground state $|0\rangle$. The first few such states are given by

$$|0\rangle \qquad a_{-1}|0\rangle \qquad a_{-2}|0\rangle, a_{-1}^2|0\rangle \qquad a_{-3}|0\rangle, a_{-2}a_{-1}|0\rangle, a_{-1}^3|0\rangle. \qquad (11.24)$$

We have grouped these states such that within each group the sum of mode numbers of the creation operators is the same. Let $N = \sum n N_n$ be the mode number operator that determines the total mode number of a given state. Then we introduce a counting function $Z(q)$ through

$$Z^B(q) = \sum_{m=0}^{\infty} (\text{number of states with } N = m) \; q^m = 1 + q + 2q^2 + 3q^3 + \cdots .$$

The first few terms we have spelled out count the states we listed in equation (11.24). It is not difficult to find an explicit formula for $Z^B(q)$. In fact, a moment of thought reveals that it should take the form

$$Z^B(q) = \prod_{n=1}^{\infty} \frac{1}{1 - q^n} =: \frac{q^{1/24}}{\eta(q)}. \qquad (11.25)$$

In order to verify that the infinite product gives the desired counting function $Z^B(q)$, one has to expand each of the factors $1/(1 - q^n) = \sum_k q^{nk}$ in a geometric series. In the previous formula we have also introduced the function $\eta(q)$ that is known in the mathematical literature as DEDEKIND's η function.

Let us now switch to the fermions. Counting the states created by the modes b_n or b_r of a single fermionic field is rather easy because any given mode can at most appear once. We treat the NS-sector first. The lowest states are given by

$$|0\rangle \;, \; b_{-1/2}|0\rangle \;, \; b_{-3/2}|0\rangle \;, \; b_{-3/2}b_{-1/2}|0\rangle \;, \ldots .$$

In this case the counting function is easily seen to take the form

$$Z_{NS}^F(q) = \prod_{n=1}^{\infty} (1 + q^{n-1/2}) = 1 + q^{1/2} + q^{3/2} + q^2 + \cdots .$$

The factors $(1 + q^{n-1/2})$ in this infinite product correspond to the fermionic modes $b_{1/2-n}$. The corresponding expression for the R-sector takes the form

$$Z_R^F(q) = \prod_{n=1}^{\infty} (1 + q^n).$$

Z_R^F counts the states in the R-sector up to a constant multiplicity that is associated with the dimension of the space of ground states.

After this preparation we can come back to our open string theory. In this case, we have eight bosonic fields, one for each transverse direction, and the same number of fermionic ones. Furthermore, the space of ground states in the R-sector has dimension 8. All this is independent of the brane's dimension $p + 1$. Putting things together, the physical states of the NS-sector are counted by

$$Z_{NS}(q) = \frac{q^{-\frac{1}{2}}}{2} \left[\prod_{n=1}^{\infty} \left(\frac{1 + q^{n-\frac{1}{2}}}{1 - q^n} \right)^8 - \prod_{n=1}^{\infty} \left(\frac{1 - q^{n-\frac{1}{2}}}{1 - q^n} \right)^8 \right]$$

$$= \frac{1}{\eta^8(q)} \left[f_3^8(q^{\frac{1}{2}}) - f_4^8(q^{\frac{1}{2}}) \right]. \tag{11.26}$$

The first term in square brackets counts all the states that are generated by the eight sets of fermionic and bosonic creation operators. All terms possess a positive coefficient. In the second term in square brackets we inverted the sign in the numerator. Otherwise the expression is identical to the first term. The resulting expression contains summands with both positive and negative coefficients. The former are related to states that contain an even number of fermionic creation operators, while the latter keep track of states with an odd number of fermionic creators. Since we subtract the two expressions, our formula counts all the states on the NS-sector that survive the GSO projection. The factor $q^{-1/2}$ ensures that massless states in the GSO projected NS-sector come with a factor q^0.

In the last step we have introduced two important functions f_3 and f_4. Along with another function f_2 that we shall use below, these special functions are defined through as follows:

$$f_2(q^{\frac{1}{2}}) = \sqrt{2} q^{\frac{1}{24}} \prod_{n=1}^{\infty} (1 + q^n) = \sqrt{\frac{\theta_2(q)}{\eta(q)}}, \tag{11.27}$$

$$f_3(q^{\frac{1}{2}}) = q^{-\frac{1}{48}} \prod_{n=1}^{\infty} (1 + q^{n-\frac{1}{2}}) = \sqrt{\frac{\theta_3(q)}{\eta(q)}}, \tag{11.28}$$

$$f_4(q^{\frac{1}{2}}) = q^{-\frac{1}{48}} \prod_{n=1}^{\infty} (1 - q^{n-\frac{1}{2}}) = \sqrt{\frac{\theta_4(q)}{\eta(q)}}. \tag{11.29}$$

All three functions are well known from the theory of θ functions. We shall discuss some of their properties below.

The counting of states in the R-sector is even simpler. In this case, the number of R-sector states is encoded in

$$Z_R(q) = 8 \prod_{n=1}^{\infty} \left(\frac{1+q^n}{1-q^n} \right)^8 = \frac{1}{\eta^8(q)} f_2^8(q^{\frac{1}{2}}). \tag{11.30}$$

The factor 8 arises from the eight ground states in the R-sector. All excited states are counted by the infinite product. In the second equality, we have expressed the product through the function f_2 that was introduced above.

Supersymmetry of the spectrum implies that the number of states in the R- and NS-sector coincide level by level. In other words, supersymmetry would predict vanishing of the quantity

$$Z(q) := Z_{NS}(q) - Z_R(q) = \frac{1}{\eta^8(q)} \left[f_3^8(q^{\frac{1}{2}}) - f_4^8(q^{\frac{1}{2}}) - f_2^8(q^{\frac{1}{2}}) \right].$$

For the first few levels one may check by hand that $Z(q)$ does indeed vanish. But in order to really prove $Z(q) = 0$ one needs some help from the theory of θ functions. In fact, searching one of the standard references on the subject (see, e.g., [46]), one can find the following "abstruse identity"[3] for the JACOBI θ functions:

$$f_3^8(q^{\frac{1}{2}}) - f_2^8(q^{\frac{1}{2}}) - f_4^8(q^{\frac{1}{2}}) = 0. \tag{11.31}$$

This implies that $Z = 0$, so that each mass level of the spectrum has the same number of bosons and fermions. Thereby, we have demonstrated that the spectrum of open string modes on a Dp-brane is supersymmetric.

11.4 10-Dimensional N=1 Super-YANG-MILLS Theory

Before we end this chapter, let us briefly talk about the low-energy effective field theory on a stack of M Dp-branes. We know from our earlier discussion in Chapter 7 what to expect: the bosonic content of the theory should be given by a U(M) gauge field and a number of scalars Φ_a, one for each of the $9 - p$ directions transverse to the brane. These scalars transform in the adjoint of the gauge group. In the supersymmetric theory, each of the bosonic fields is joined by a fermionic partner. Since a massless gauge boson on the brane has $p - 1$ physical degrees of freedom, the total number $9 - p + p - 1 = 8$ of bosons and fermions is independent of the brane's dimension. In 10 dimensions, 8 bosons and fermions form a massless multiplet of the N=1 POINCARÉ superalgebra; see Chapter 9. The

---

[3] In the literature also known as the JACOBI identity.

gauge theories for lower dimensional branes can be obtained from the N=1 super-YANG-MILLS theory in a 10-dimensional space-time by a process called dimensional reduction, i.e., by setting $\partial_a F = 0$ for $a = p + 1, \ldots, 9$ and all fields F in the theory.

The action of N=1 super-YANG-MILLS theory in 10 dimensions contains two dynamical objects: a gauge field A and a MAJORANA-WEYL fermion ψ. Being the superpartner of the matrix valued gauge fields A, ψ must also take values in the space of $M \times M$ matrices. The action is

$$S[A, \psi] = \int d^{10}x \; \mathrm{tr} \left(-\frac{1}{4}F^2 + \frac{i}{2}\bar{\psi}\gamma_\mu D^\mu \psi \right). \tag{11.32}$$

Here, F^2 denotes the square $F_{\mu\nu}F^{\mu\nu}$ of the YANG-MILLS field strength. The latter is obtained from the gauge field through the usual formula:

$$F_{\mu\nu} = \partial_\mu A_\nu - \partial_\nu A_\mu + g[A_\mu, A_\nu].$$

The covariant derivative D_μ in the kinetic term for the fermions takes the following form:

$$D_\mu \psi = \partial_\mu \psi + g[A_\mu, \psi].$$

The field ψ contains all 16 components of a 10-dimensional MAJORANA-WEYL fermion. In a slight abuse of notation we have used the symbols γ^μ to denote the Dirac matrices projected to the 16-dimensional WEYL representation of the LORENTZ algebra.

The action (11.32) is the natural supersymmetric extension of a 10-dimensional YANG-MILLS theory, and it possesses the right amount of supersymmetry to describe the massless open string modes in the case where we impose NEUMANN boundary conditions in all 10 directions. After passing to Dp-brane boundary conditions, the bosonic zero modes $p^a, a = p + 1, \ldots, 9$ vanish, $p^a = 0$. Otherwise, the theory does not depend on the dimension p of the Dp-brane. In particular, the number of supercharges is always 16, regardless of the value p takes. The corresponding low-energy effective action can be determined simply by setting $\partial_a = 0$ for $a = p + 1, \ldots, 9$.

As an example, consider a stack of D3-branes. Through dimensional reduction we obtain a 4-dimensional field theory with 16 supercharges. Since MAJORANA fermions in four dimensions contain four components, the 16 supercharges of the dimensionally reduced theory are the fermionic generators of an N=4 POINCARÉ superalgebra in four dimensions. In order to describe the action of the N=4 supersymmetric gauge theory, we separate the gauge field of the 10-dimensional model as $A_\mu = (A_\alpha, \phi_a)$ with $\alpha = 0, 1, 2, 3$ and $a = 4, \ldots, 9$, i.e., we split A_μ into the gauge boson A_α and six scalars ϕ_a of the 4-dimensional model. Dimensional reduction leads to the following action:

$$S = \int d^4x \ \mathrm{tr} \left(-\frac{1}{4}F^2 - \frac{1}{4}D_\alpha\phi_a D^\alpha\phi_a - \frac{g^2}{4}[\phi_a, \phi_b]^2 + \text{fermionic terms} \right).$$

$$(11.33)$$

The three terms we have displayed here are obtained from the $F_{\mu\nu}F^{\mu\nu}$ in our N=1 super-YANG-MILLS theory (11.32). The YANG-MILLS action in four dimensions collects all terms with both $\mu, \nu = 0, 1, 2, 3$. Terms with $\mu = 0, 1, 2, 3$ and $\nu = 4, \dots, 9$ (and vice versa) give rise to the kinetic terms for the six scalar fields ϕ_a. Their fourth order interaction, finally, emerges from the 10-dimensional YANG-MILLS action when both μ and ν take values in the set $\{4, \dots, 9\}$. The N=4 super-YANG-MILLS theory (11.33) is a very important field theory. Over the last decade the investigation of this model has contributed a great deal toward developing new methods for 4-dimensional quantum field theories. We shall see a bit of this later, in our discussion of the AdS/CFT correspondence.

12

Construction of Heterotic Superstrings

So far we have discussed two string theories with 32 supercharges, i.e., with N=2 supersymmetry in 10 dimensions. One may wonder whether there are any theories that possess the minimal number of 16 supercharges. We shall see that this is the case, and we want to describe the corresponding models both in the supergravity approximation and in string theory. What we will find are two consistent closed string theories with a tachyon free spectrum and N=1 supersymmetry. These are known as heterotic SO(32) and $E_8 \times E_8$ theory and were discovered by GROSS, HARVEY, MARTINEC, and ROHM in [33].

12.1 *10*-Dimensional N=1 Supergravity Theories

It is useful to begin with a short discussion of N=1 supermultiplets in 10 dimensions. Recall that the 10-dimensional POINCARÉ superalgebra has 16 supercharges in one of the 16-dimensional spinor representations **16** or **16**$'$ that we have constructed in Chapter 9. These supercharges obey the following anti-commutation relations:

$$\left\{ Q_a, \overline{Q}_b \right\} = -2P_\mu \left(\Pi_\pm \Gamma^\mu \right)_{ab}. \tag{12.1}$$

Our notations are the same as in Chapter 9. We are interested in massless multiplets of this algebra. According to our general theory, these may be constructed on a state space of the form $L_2(\{p^2 = 0\}|V)$ where the finite dimensional vector space V carries an irreducible action of so(8) along with the action of Q_a subject to the relations

$$\left\{ Q_a, \overline{Q}_b \right\} = -2p_0^\mu \left(\Pi_\pm \Gamma_\mu \right)_{ab}. \tag{12.2}$$

Here, the vector p_0^μ denotes a point on the mass-hyperboloid $p^2 = -m^2$. For massless multiplets, for example, our standard choice is $p_0 = (1, 1, 0, \ldots, 0)$.

A study of the relations (12.2) shows that eight of the generators Q anti-commute with all the other generators, and hence they can be set to zero. The other eight generators may then be combined into four creation and four annihilation operators. Hence the dimension of the finite dimensional vector space V for massless multiplets is $dim\ V = dim\ V_0 \cdot 2^4$.

There are two types of multiplets to discuss here. The first one is the N=1 *vector multiplet* that we have seen before. It is obtained by acting with creation operators on the 1-dimensional trivial representation of so(8). Under the action of so(8), the resulting 16-dimensional space V decomposes into two multiplets:

$$V|_{so(8)} \simeq \mathbf{8}_V \oplus \mathbf{8}_{S/C}. \tag{12.3}$$

The eight bosonic states correspond to eight physical polarizations of a vector boson in 10 dimensions. In addition there are eight fermionic degrees of freedom. There is yet another multiplet that will become important for us, namely the *graviton multiplet*. It is constructed by acting with four creation operators on ground states in an 8-dimensional vector representation of so(8). What we obtain is a 128-dimensional space V that decomposes into the following irreducible representations of so(8):

$$V|_{so(8)} \simeq \underbrace{\mathbf{1}}_{\text{dilaton}} \oplus \underbrace{\mathbf{28}}_{\text{KALB-RAMOND}} \oplus \underbrace{\mathbf{35}}_{\text{graviton}} \oplus \mathbf{8}_{S/C} \oplus \mathbf{56}_{S/C}. \tag{12.4}$$

We shall employ one copy of the graviton supermultiplet and a certain number of vector supermultiplets to build N=1 supergravity theories,

$$S \sim \frac{1}{2\kappa_{10}^2} \int d^{10}x \sqrt{-G} e^{-2\phi} \left[\mathcal{R} + 4(\partial\phi)^2 - \frac{1}{2}(\tilde{H}_3)^2 + \frac{\kappa_{10}^2}{g_{10}^2} \mathrm{tr}(|F_2^2|) \right], \tag{12.5}$$

where F_2 denotes the field strength of a YANG-MILLS gauge field A_μ for a (possibly non-Abelian) gauge group G. The action for the graviton multiplet is the familiar one, except that we slightly modified the 3-form $H_3 = dB$ by incorporating a so-called CHERN-SIMONS 3-form of the gauge field:

$$\tilde{H}_3 = dB + \frac{\kappa_{10}^2}{g_{10}^2}\omega_3 \quad \text{where} \quad \omega_3 = \mathrm{tr}\left(A \wedge dA + \frac{2}{3}A \wedge A \wedge A\right). \tag{12.6}$$

After such a shift, the gauge and gravity sector of this N=1 supergravity theory are coupled through the term $|\tilde{H}_3|^2$.

All these classical actions possess diffeomorphism symmetry, gauge symmetry, and N=1 supersymmetry. The trouble starts with the quantization as classical symmetries are often broken by quantum effects. In the case of gauge symmetries, such anomalies in the quantum theory can be disastrous, i.e., they can spoil consistency of the model. We have seen one such example at the beginning of this book when we tried to quantize bosonic field theory such that the classical re-parametrization symmetry could be fixed in the quantum theory. Ultimately, we had to choose $D = 26$ in order to obtain a consistent theory with no negative norm states. A similar analysis of quantum consistency can be performed for the N=1 supergravities we have described above. One loop computations show that the spectrum will contain negative norm states unless we make very special choices for the YANG-MILLS field. In fact, the theory can be shown to be free of anomalies only when the gauge fields take values in either so(32) or the product algebra $E_8 \times E_8$; see [29].

The 496-dimensional Lie algebra so(32) does not need much explanation. On the other hand, we shall need a little bit of additional background on E_8. The Lie algebra E_8 contains the 120-dimensional subalgebra so(16) $\subset E_8$. All other generators of E_8 transform in a single representation of so(16), namely in the 128-dimensional spinor representation of so(16). Hence the total number of generators for E_8 is $dim\ E_8 = 248$. We shall denote the generators of so(16) by $T^{\alpha\beta}, \alpha, \beta = 1, \ldots, 16$. The 128 generators in the spinor representation are denoted by $U^a, a = 1, \ldots, 128$. With these notations we can now spell out the relations of E_8:

$$[T^{\alpha\beta}, T^{\gamma\sigma}] = i\delta_{\beta\sigma} T^{\alpha\gamma} - i\delta_{\beta\gamma} T^{\alpha\sigma} - i\delta_{\alpha\sigma} T^{\beta\gamma} + i\delta_{\alpha\gamma} T^{\beta\sigma}, \tag{12.7}$$

$$[T^{\alpha\beta}, U^a] = -\frac{i}{4} [\gamma^\alpha, \gamma^\beta]^a{}_b U^b, \tag{12.8}$$

$$[U^a, U^b] = -i \gamma^{[\alpha}_{ac} \gamma^{\beta]}_{cb} T^{\alpha\beta}. \tag{12.9}$$

Note that the commutation relations for the generators U^a and U^b involve the anti-symmetric Lie bracket. So, E_8 is an ordinary Lie algebra and not a supersymmetric extension of so(16). One may show that the bracket we have specified obeys the usual JACOBI identity.

12.2 Heterotic String Theory: The Idea

If we look back into our discussion of type II string theories, we can easily locate the reason we obtained N=2 supersymmetry. In fact, our analysis first focused on the right-moving sector where, after GSO projection, we found massless modes in the representation $8_V \oplus 8_{S/C}$ of so(8). Now we recognize this as the content of a massless vector multiplet for the N=1 POINCARÉ superalgebra. Then we multiplied this result from the right-moving degrees of freedom with a similar contribution from the left movers and obtained the spectrum with N=2 supersymmetry. The natural idea for the construction of spectra with N=1 supersymmetry therefore is to break the symmetry between the left and right movers. The simplest way in which N=1 supersymmetry could emerge would be to combine an N=1 supersymmetric spectrum for the right movers with a non-suspersymmetric spectrum for the left movers. This is the main idea of heterotic string theories.

Throughout this book we have learned how to construct supersymmetric and non-symmetric spectra in string theory. Let us briefly review the two main constructions we have used. In bosonic string theory we worked with a 2-dimensional auxiliary field theory that contained the generators L_n of some VIRASORO algebra with $c = 26$. Imposing the VIRASORO constraints

$$L_0|\psi\rangle = |\psi\rangle, \quad L_n|\psi\rangle = 0 \quad \text{for } n > 0,$$

we obtained a state space without negative norm states. The spectrum possesses POINCARÉ symmetry without any supersymmetry. A very similar scenario will later be used in the left-moving sector of heterotic string theory.

The second construction we used resulted in a supersymmetric spectrum. Here we worked with an auxiliary field theory that contained the generators L_n and $G_{n/r}$ of some N=1 super-VIRASORO algebra of central charge $c = 10 + 10/2 = 15$. Imposing the super-VIRASORO constraints

$$L_0|\psi\rangle_{NS} = \frac{1}{2}|\psi\rangle_{NS}, \quad L_n|\psi\rangle_{NS} = 0 = G_r|\psi\rangle_{NS} \quad \text{for } n, r > 0,$$

$$L_0|\psi\rangle_R = \frac{5}{8}|\psi\rangle_R, \quad L_n|\psi\rangle_R = 0 = G_n|\psi\rangle_R \quad \text{for } n > 0,$$

we obtained a state space without negative norm states and super-POINCARÉ symmetry. In all cases, the operator M^2 that was measuring the mass of the multiplets of the spectrum can be written as

$$M^2 = \frac{4}{\alpha'}\left(-\bar{a}^* + \bar{L}'_0\right) = \frac{4}{\alpha'}\left(-a^* + L'_0\right), \tag{12.10}$$

where L'_0 and $\bar{L}'_0$ denote the VIRASORO generators without contributions from the zero modes. The value of the number a^* (and $\bar{a}^*$) depends on the type of the construction and on the sector. In case we are dealing with the usual VIRASORO constraints, $a^* = a^V$ is chosen to be $a^V = 1$. In the case of super-VIRASORO constraints, the value of a^* depends on whether we are in the R- or the NS-sector of the theory. In the R-sector $a^* = a_R^{sV}$ takes the value $a_R^{sV} = 5/8$. For states in the NS-sector, the value of $a^* = a_{NS}^{sV}$ is given by $a_{NS}^{sV} = 1/2$. In order to make the formula for M^2 more explicit, we express L'_0 and $\bar{L}'_0$ in terms of oscillator counting operators $N_{n/2}$:

$$L'_0 = \sum_{n=1}^{\infty} \frac{n}{2} N_{n/2} + \frac{n_P^f}{16}, \quad \bar{L}'_0 = \sum_{n=1}^{\infty} \frac{n}{2} \bar{N}_{n/2} + \frac{\bar{n}_P^f}{16}.$$

The numbers n_P^f and $\bar{n}_P^f$ count the right- and left-moving fermions with periodic boundary conditions. The necessity of such a constant shift was argued for in Chapter 8. Putting all this together, we arrive at the following expression for the mass operators:

$$M^2 = \frac{4}{\alpha'}\left(\kappa + \sum_{n=1}^{\infty}\frac{n}{2}N_{n/2}\right) = \frac{4}{\alpha'}\left(\bar{\kappa} + \sum_{n=1}^{\infty}\frac{n}{2}\bar{N}_{n/2}\right), \qquad (12.11)$$

where

$$\kappa = -a^* + \frac{n_P^f}{16}, \qquad \bar{\kappa} = -\bar{a}^* + \frac{\bar{n}_P^f}{16}$$

denote the lowest eigenvalue of M^2 in the left and right sector, respectively, before GSO projection. For later reference, let us also repeat the values of a^* (and $\bar{a}^*$) that can appear:

$$a^V = 1, \quad a_R^{sV} = \frac{5}{8}, \quad a_{NS}^{sV} = \frac{1}{2}. \qquad (12.12)$$

It might be instructive to reproduce our previous results from these formulas. If we are in the right-moving R-sector of a theory with 10 bosonic and 10 fermionic fields, for example, then $n_P^f = 10$ and $a^* = a_R^{sV} = -5/8$ so that $\kappa = -5/8 + 10/16 = 0$, in agreement with our analysis in Chapter 10.

After this preparation, let us proceed to build the class of new theories we are interested in. Since we want a theory of strings that propagate in 10-dimensional MINKOWSKI space, we will include 10 bosonic fields X^μ. These contribute $c^b = 10 = \bar{c}^b$ to both the left- and right-moving central charge. Our aim is to build a model that has left central charge $\bar{c} = 26$, while we want the right central to be $c = 15$. What we did in bosonic string theory was to add 16 additional bosonic fields. But this brought both the left and the right central charge up to 26, and hence it is not an option for us. Here is where fermionic fields come in very handy. Recall that 2-dimensional MAJORANA fermions possess two components, one of which is purely right-moving while the other is purely left-moving. These two components do not interact and hence can be put into the model separately. Each left-moving fermion λ^α contributes an amount $1/2$ to the left central charge, and hence we need 32 of them in order to reach $\bar{c} = 26$. At the same time, we can add 10 right-moving fermions ψ^μ in order to bring the right central charge to the desired value $c = 15$. These arguments motivate to look at class of models with action

$$S^{FF}[X, \psi_+, \lambda] = \frac{1}{\pi\alpha'}\int d\tau d\sigma \, \partial_+ X^\mu \partial_- X_\mu + \frac{1}{2\pi}\int d\tau d\sigma \, \left(\lambda^\alpha \partial_- \lambda_\alpha + \psi_-^\mu \partial_+ \psi_{-,\mu}\right), \qquad (12.13)$$

Table 12.1 *The values of the constants $\bar{\kappa}$ and κ in the various sectors of the so(32) symmetric theory before GSO projection.*

$(\bar{\kappa}, \kappa)$	R	NS
P	$(+1, 0)$	$\left(+1, -\dfrac{1}{2}\right)$
A	$(-1, 0)$	$\left(-1, -\dfrac{1}{2}\right)$

where $\alpha = 1, \ldots, 32$. From the right-moving fields in this theory we can build the generators of an N=1 superconformal algebra as in Chapter 8. In the left sector we can use the modes a_n^μ of the bosonic fields along with modes $c_{n/r}^\alpha$ of the fermionic fields λ^α in the R- and NS-sector to build the generators L_n of a VIRASORO algebra with $\bar{c} = 26$.

12.3 SO(32) Heterotic String Theory

From the class of theories that are described by the action (12.13) we can build several string theories that differ from each other by the choice of boundary conditions (and GSO projections) for the left-moving fermions λ^α. Our first possibility is to treat all these 32 fermions in the same way, i.e., impose the same boundary conditions. Hence, we obtain two sectors, one when all 32 fermions satisfy periodic (P) boundary conditions, the other if they are simultaneously subject to anti-periodic (A) boundary conditions. By construction, a model of this type will possess an so(32) symmetry.

In order to understand the mass spectrum of such a theory, we first need to compute the constants κ and $\bar{\kappa}$ for all four sectors (P, R), (P, NS), (A, R), and (A, NS) of the model. The results are summarized in Table 12.1. The values $\kappa_{NS} = -1/2$ and $\kappa_R = 0$ for the constant κ in the right sector are well known to us. In the left sector with periodic boundary conditions on all fermions, $\bar{\kappa}_P = -1 + 32/16 = 1$. When no fermion satisfies periodic boundary conditions (because they are all antiperiodic), then we obtain $\bar{\kappa}_A = -1$. Constructing the spectrum of heterotic string theory also involves a GSO projection. For the right sector, the corresponding projection operators are the same as in Chapter 10; see discussion around eq. (10.12). On the left side, we propose to keep the ground state of the anti-periodic sector in the spectrum, i.e., in the sector A we shall use

$$\overline{\Pi}^A = \frac{1}{2}\left(1 + \overline{\Gamma}_\pm^A\right).$$

Here, following the first part of eq. (10.12), $\overline{\Gamma}^A_\pm$ is defined such that it anti-commutes with the modes c^α_r of the fermions λ^α, and it is -1 on the ground state of the anti-periodic sector. A pair of similar operators $\overline{\Pi}^P_\pm$ is used in case of periodic boundary conditions. As their right-moving counterpart $\Pi^R_\pm$, these operators involve a product of fermionic zero modes.

For a state of the field theory to describe a wave function of a closed string, the eigenvalues of the left- and right-moving operator M^2 must be identical. This implies that our spectrum is free of tachyons simply because – after the GSO projection with Π^{NS} – the mass operator satisfies $M^2 \geq 0$ in the right sector of the model. So, let us look at the massless modes. Obviously, the massless spectrum of the theory does not receive any contribution from the sectors (P, R) and (P, NS) simply because $M^2 \geq 1$ on the left side of the theory.

The remaining two sectors do contribute to the massless spectrum. Space-time bosons come from the $\mathbf{8}_V$ in the right NS-sector, while fermions arise from the $\mathbf{8}_{S/C}$ in the right R-sector. These are both multiplied with the same contribution $(\mathbf{8}_V, \mathbf{1})$ from the anti-periodic left sector. The result, including contributions from the periodic left sector, takes the form

$$\underbrace{[(\mathbf{8}_V, \mathbf{1}) \oplus (\mathbf{1}, \mathbf{496})]}_{\text{left movers}} \times \underbrace{\left[(\mathbf{8}_V, \mathbf{1}) \oplus (\mathbf{8}_{S/C}, \mathbf{1}) \right]}_{\text{right movers}} \qquad (12.14)$$

$$= \underbrace{(\mathbf{1}, \mathbf{1})}_{\text{dilaton}} \oplus \underbrace{(\mathbf{28}, \mathbf{1})}_{\text{KALB-RAMOND}} \oplus \underbrace{(\mathbf{35}, \mathbf{1})}_{\text{graviton}} \oplus \underbrace{(\mathbf{8}_V, \mathbf{496})}_{\text{so(32)–vector}} \qquad (12.15)$$

$$+ \text{ fermionic partners.}$$

Here we have listed each multiplet (a, b) with two labels. While the first one refers to the so(8) symmetry, the second one keeps track of the so(32) gauge symmetry. Since quantities in the right-moving sector do not carry any gauge quantum numbers, the second entry is trivial. As for the left sector, our reasoning goes as follows. Let us first note that the lowest eigenvalue of M^2 is $\bar{k} = -1$. Hence, the action of creation operators has to add one unit of M^2 in order to match the value of M^2 in the right sector. There are two different ways to do so. One way is to act with the modes a^μ_{-1} on the ground state $|k\rangle \otimes |A\rangle$. Since our GSO projection keeps $|A\rangle$ in the spectrum, such a state is not projected out by the GSO projection. After imposing physical state conditions and removing the zero norm state we remain with the $\mathbf{8}_V$ of so(8). These states transform trivially under the so(32) gauge symmetry. But there is another way to obtain states with $M^2 = 0$. In fact, we can also act with two fermionic modes $c^\alpha_{-1/2}$. The resulting states

$$c^\alpha_{-1/2} c^\beta_{-1/2} |k\rangle \otimes |A\rangle$$

Table 12.2 *The values of the constants $\bar{\kappa}$ and κ in the various sectors of the $E_8 \times E_8$ symmetric theory before GSO projection.*

$(\bar{\kappa}, \kappa)$	R	NS
(P, P)	$(+1, 0)$	$\left(+1, -\dfrac{1}{2}\right)$
(A, P)	$(0, 0)$	$\left(0, -\dfrac{1}{2}\right)$
(P, A)	$(0, 0)$	$\left(0, -\dfrac{1}{2}\right)$
(A, A)	$(-1, 0)$	$\left(-1, -\dfrac{1}{2}\right)$

are not removed by the GSO projection since they contain an even number of fermionic creation operators. Thereby, we obtain $32 \cdot 31/2 = 494$ states in the 496-dimensional adjoint representation of so(32). After multiplying the content of the left and right sector, we end up with the massless spectrum of the N=1 supergravity with gauge group SO(32).

12.4 $E_8 \times E_8$ Heterotic String Theory

Our second model arises once we split the 32 fermions λ^α into two groups of 16 each. Let us denote these fermionic fields by λ_1^α and λ_2^α with $\alpha = 0, \ldots, 16$. In this case, the left movers of the model decompose into four sectors, which we shall label by $(P, P), (A, P), (P, A)$, and (A, A). Table 12.2 displays the value of $(\bar{\kappa}, \kappa)$ in all the different sectors.

The only new value of $\bar{\kappa}$ arises in the two sectors (A, P) and (P, A), which both contain 16 fermions with periodic boundary conditions. Hence, we find $\bar{\kappa} = -1 + 16/16 = 0$. Next we need to perform the GSO projection. We do this separately for the two groups of left-moving fermions. In the (A, A) sector of the model, for example, the left GSO projection takes the form

$$\overline{\Pi}^{(A,A)} = \overline{\Pi}_1^A \overline{\Pi}_2^A \quad \text{where} \quad \overline{\Pi}_i^A = \frac{1}{2}\left(1 - \overline{\Gamma}_i^A\right).$$

$\overline{\Gamma}_1^A$ anti-commute with the first 16 fermions and commute with the remaining 16. For $\overline{\Gamma}_2^A$ it is the other way around. Both operators have eigenvalue -1 on the ground state. This left GSO projection keeps the ground state $|k\rangle \otimes |A\rangle_1 \otimes |A\rangle_2$ in the theory. The resulting GSO projected state space is clearly tachyon free, for the same reason as in the previous section. The massless spectrum does receive contributions from all sectors except for the (P, P) sector in which $M^2 \geq 1$. For the sector (A, A) we find

$$[(\mathbf{8}_V, \mathbf{1}, \mathbf{1}) \oplus (\mathbf{1}, \mathbf{1}, \mathbf{120}) \oplus (\mathbf{1}, \mathbf{120}, \mathbf{1})] \otimes [\mathbf{8}_V \oplus \mathbf{8}_{S/C}] \tag{12.16}$$

$$= \underbrace{(\mathbf{1}, \mathbf{1}, \mathbf{1})}_{\text{dilaton}} \oplus \underbrace{(\mathbf{28}, \mathbf{1}, \mathbf{1})}_{\text{KALB-RAMOND}} \oplus \underbrace{(\mathbf{35}, \mathbf{1}, \mathbf{1})}_{\text{graviton}} \oplus (\mathbf{8}_V, \mathbf{120}, \mathbf{1}) \oplus (\mathbf{8}_V, \mathbf{1}, \mathbf{120}) \tag{12.17}$$

$+$ fermionic partners.

The reasoning can be copied from our discussion in the previous section, except that now both of the fermionic creation operators need to be chosen from the same subset of 16 fields in order to survive the GSO projection. This leaves us with $16 \cdot 15/2 = 120$ states that transform in the adjoint representation of the two SO(16) factors in the gauge group $\mathrm{SO}(16)_1 \times \mathrm{SO}(16)_2$. Turning to the sectors (A, P) and (P, A) we find

$$[(\mathbf{1}, \mathbf{128}, \mathbf{1}) \oplus (\mathbf{1}, \mathbf{1}, \mathbf{128})] \otimes [\mathbf{8}_V \oplus \mathbf{8}_{S/C}] \tag{12.18}$$

$$= (\mathbf{8}_V, \mathbf{128}, \mathbf{1}) \oplus (\mathbf{8}_V, \mathbf{1}, \mathbf{128}) \oplus \text{fermionic partners.} \tag{12.19}$$

The 128 states survive the GSO projection on the ground states of 16 fermions with periodic boundary conditions. In fact, the fermions λ_i^α possess 16 zero modes each. As before, we form eight creation operators from the 16 zero modes and build a space of ground states of dimension 2^8. Half of them survive the GSO projection. These 128 states transform in the spinor representation of so(16).

Putting the last two formulas together, we obtain the particle content of N=1 supergravity with gauge group $E_8 \times E_8$. In particular, the multiplets $(\mathbf{8}_V, \mathbf{120}, \mathbf{1})$ and $(\mathbf{8}_V, \mathbf{128}, \mathbf{1})$ of so(8) $\otimes$ so(16) $\otimes$ so(16) combine in an $(\mathbf{8}_V, \mathbf{248}, \mathbf{1})$ of so(8) $\otimes E_8 \otimes E_8$, i.e., into the vector multiplet for gauge group $E_8 \times E_8$.

12.5 Concluding Remarks

We conclude this introduction to heterotic string theory with a few short remarks. One may certainly wonder whether we could have continued our analysis by splitting the 32 left-moving fermions into smaller and smaller groups. This could potentially lead to N=1 theories with different gauge groups. On the other hand, the corresponding N=1 supergravities are known to possess anomalies. The conflict can easily be resolved: It turns out that all other candidate state spaces for additional heterotic string theories contain tachyonic modes. Hence, the string theoretic constructions reproduce only anomaly free supergravities, as they are supposed to.

After our construction of type II theories we had gone on to discuss associated theories with Dp-brane boundary conditions. One might be tempted to look for an extension of that analysis to heterotic models. N=1 supergravity theories do possess classical solutions that source the gauge fields in the graviton multiplet, i.e., the KALB-RAMOND field. These are the N=1 analogues of the fundamental string and NS5 brane solutions that we discussed briefly when we studied branes in type II theories. On the other hand, the theories of open strings we constructed in Chapter 11 modeled a different type of solution, namely those that were sourcing the RR-fields of type II supergravity. Such solutions do not exist in the heterotic models simply because there are no RR-fields. Looking at our string theory construction we come to the same conclusions. Recall that the construction of Dp-brane boundary conditions in type II theories required identifying the left- and right-moving modes along the boundary. Such an identification is not possible in heterotic theories since the left- and right-moving sectors possess a different field content. Therefore, there are no generalizations of D-branes in heterotic string models.

Let us finally mention that there is one more consistent string theory, known as type I string theory, whose low-energy effective field theory is given by the N=1 SO(32) supergravity. It descends from type IIB theory through an appropriate identification of left- and right-moving modes. This removes most of the form fields of the type IIB theory. In fact, only the RR 2-form field C_2 survives. The resulting theory of unoriented closed strings is not consistent unless 32 space-time filling branes are introduced. We end up with the string theory whose low-energy effective action is given by eq. (12.13), except that B is to be replaced by the RR-field C_2. The construction of type I theory requires a number of new elements and is beyond the scope of these chapters; see chapter 10 of [56] for an introduction and references.

Part II

Strings in Curved Backgrounds

In the first part of this book we have discussed (super-)string theory in flat 26(10)-dimensional backgrounds. It is clear that such string theories are phenomenologically not relevant. Our main task in the second part is to understand some of the technology that is used to build string theories that are relevant, in particular for 4-dimensional physics. The basic idea is to study the motion of superstrings on backgrounds in which some of the dimensions are curled up. This process is known as compactification. As we shall discuss, describing strings in curved backgrounds is a highly non-trivial task. We shall sketch a few of the basic technologies in the next three chapters. This will be followed by two more blocks of chapters, one on CALABI-YAU compactifications, the other on string dualities, with special emphasis on the AdS/CFT correspondence and its application to gauge theory.

13

Introduction to Conformal Field Theory

The purpose of this chapter is to make the transition from the 2-dimensional free field theories we studied extensively in the first part of this book to those theories that are relevant for string theory in backgrounds with a non-constant metric. The latter will require us to re-think the way in which we approach 2-dimensional (conformal) quantum field theories. As we shall see, a central role is played by infinite dimensional symmetries. So, we have to talk about representation theory, characters, and modular transformations. In addition, we shall explore the construction of correlation functions and discover that the free field theories we worked with up to now are just special members within a continuous family of local field theories.

13.1 Compactification and CFT Backgrounds

There are a number of good reasons to study strings in curved backgrounds. To begin, our world does not look at all like a flat 10-dimensional MINKOWSKI space. Therefore at least some 10 directions must be compactified, i.e., curled up into a small compact space. One class of such compactifications that has been studied extensively involves backgrounds of the form

$$\mathcal{X}^{1,9} \sim \underbrace{\mathcal{X}^{1,3}}_{\text{cosmological}} \times \underbrace{\mathscr{X}^6}_{\text{compact}} , \qquad (13.1)$$

where $\mathcal{X}^{1,3}$ is some non-compact cosmological background and $\mathscr{X}^6$ is compact. The non-compact factor $\mathcal{X}^{1,3}$ could be, e.g., a FRIEDMAN-ROBERTSON-WALKER geometry, but we shall always assume the simpler choice $\mathcal{X}^{1,3} \sim \mathcal{M}^4$ of an external MINKOWSKI space. As for the compact part $\mathscr{X}^6$, there are many possibilities, including a torus T^6, quotients T^6/G of a torus T^6 by the action of some discrete group G, CALABI-YAU manifolds, etc.

More recently, mostly within the context of gauge/string dualities, other classes of curved backgrounds have attracted considerable attention. In fact, as our discussion in the introduction suggested, it is possible to describe certain lower dimensional gauge theories through closed strings that move in the near horizon geometry of branes. The latter are always curved due to the presence of the massive brane. A key example in this class of curved space string theories is provided by the background

$$\mathcal{X}^{1,9} \sim AdS_5 \times S^5, \tag{13.2}$$

where AdS_5 denotes a 5-dimensional Anti-de Sitter space and S^5 is a 5-sphere. Closed string theories on this background are used to describe a certain 4-dimensional supersymmetric gauge theory; see Chapter 19.

Whatever motivation we have to explore strings in curved geometries, we are led to consider 2-dimensional field theory with an action of the form

$$S[X] = -\frac{1}{4\pi\alpha'} \int d\tau d\sigma \underbrace{g_{\mu\nu}(X)}_{\text{non-constant}} \partial^a X^\mu \partial_a X^\nu + \cdots \text{further terms and fermions.}$$

$$\tag{13.3}$$

Once again, this theory must be supplied with additional constraints to select the physical states of the string. There is one profound difference with the flat space case: Note that we had to contract the free indices μ and ν with the target space metric $g_{\mu\nu}$. In MINKOWSKI space, $g_{\mu\nu} = \eta_{\mu\nu}$ is constant so that the resulting action S is quadratic in the bosonic fields X^μ. But once we consider non-trivial backgrounds, $g_{\mu\nu}$ becomes a function of the coordinate fields X^μ, and hence the action is no longer quadratic. Consequently, the associated equations of motion are non-linear, and the resulting 2-dimensional field theory becomes interacting. Quantizing such models is obviously a highly non-trivial task. Recall that to date, not a single 4-dimensional interacting field theory has been quantized beyond the computation of a few leading loop orders, at least beyond the so-called planar limit.

In 2-dimensional quantum field theory, however, some very efficient techniques have been developed – starting from the seminar work [9] – to construct interacting quantum field theories, i.e., to find their spectrum and compute correlation functions. Most of these techniques are based on the use of symmetries. Everywhere in physics, symmetries are employed to simplify problems. One example we have all seen is the computation of the spectrum in the hydrogen atom. Quantum field theories, however, possess infinitely many degrees of freedom, and hence, symmetries are expected to be of significant aid only if they are infinite dimensional as well.

The quantum field theories we are interested in must admit two commuting actions of the VIRASORO generators L_n and $\bar{L}_n$. One may show that this is the case, provided that the stress energy tensor of the theory is traceless. For theories of the type (13.3), the trace of the stress energy tensor can be computed perturbatively in the coupling α'. To leading order in α' the result is (see, e.g., [27] for a derivation and references),

$$T_{-+} \sim -\alpha' R_{\mu\nu}(X) \, \partial_- X^\mu \partial_+ X^\nu + \cdots .$$

Here, $R_{\mu\nu}$ denotes the curvature tensor of the background metric $g_{\mu\nu}(X)$. Hence, for the 2-dimensional quantum field theory defined by eq. (13.3) to admit the construction of VIRASORO modes, the background metric $g_{\mu\nu}$ must be a solution of EINSTEIN equations, at least to leading order in the coupling α'. In this sense, strings select solutions of EINSTEIN equations as their natural backgrounds.

Quantum field theories that possess a traceless stress energy tensor are known as *conformal field theories*. One may consider the generators L_n and $\bar{L}_n$ of their two commuting VIRASORO algebras as generators of an infinite dimensional symmetry. In most cases, this built-in symmetry is not sufficient to solve a conformal field theory. But in many of the interesting conformal field theories, the VIRASORO symmetry gets augmented by further symmetry generators, and the resulting algebraic structure is often large enough to construct symmetry-aided solutions. We shall see one very non-trivial example a bit later. For the rest of this chapter the goal is merely to illustrate the idea of a symmetry-aided solution at the example of the free boson.

13.2 Free Field Theory Revisited: The Symmetry

In the first part of this book we have solved the free bosonic field. Now we would like to re-derive the solution, and in particular our formula for correlation functions of vertex operators from symmetries of the theory. The symmetry in question is generated by the modes a_n of the holomorphic field $J(z) = i\partial X(z, \bar{z})$ and its anti-holomorphic partner $\bar{J}$. Recall that these obey the following algebraic relations:

$$[a_n, a_m] = n \, \delta_{n+m,0}, \tag{13.4}$$

which are also known as the defining relations of a u(1) current algebra. In addition, we have the VIRASORO modes L_n and their partners $\bar{L}_n$. These are constructed from the a_n through the so-called SUGAWARA construction:

$$L_n = \frac{1}{2} \sum_{m\in\mathbb{Z}} : a_{n-m} a_m : . \tag{13.5}$$

We shall think of the modes a_n, L_n and $\bar{a}_n$, $\bar{L}_n$ as the generators of an infinite dimensional symmetry algebra. Note that this symmetry encompasses almost all degrees

of freedom of the free boson. The only operator that is not part of the symmetry is
the zero mode coordinate x.

Let us look at the algebra of a_n and L_n and study its properties as if we were
mathematicians. Whenever a mathematician is given some algebraic structure, a
natural impulse is to analyze its representations. The representations we are interested in are those in which the spectrum of the generator L_0 is bounded from below.
Since

$$[L_0, a_n] = -na_n,$$

such representations must contain at least one state of lowest weight, i.e., a state
$|\psi\rangle$ that is annihilated by all a_m with $m > 0$. The space of such lowest weight states
carries a representation of the algebra of zero modes. In our case, the only relevant
zero mode is a_0. Its algebra is Abelian, and hence the irreducible representations of
the zero mode algebra are all 1-dimensional. They are characterized by the value
k that the operator a_0 assumes, i.e., $a_0|k\rangle = k|k\rangle$. The rest of the representation
space is generated from the highest weight state $|k\rangle$ by application of the creation
operators $a_{-n}, n > 0$. Thus, the corresponding representation space of the u(1)
current algebra is given by

$$\mathcal{H}_k = \text{span}\left\{a_{-m}^{n_m} \cdots a_{-1}^{n_1}|k\rangle \, |n_i \geq 0\right\}. \tag{13.6}$$

It is not difficult to see that all these spaces $\mathcal{H}_k$ carry an irreducible representation
of the current algebra.

For any such representation we introduce a counting function, the so-called
character $\chi_k(q)$ of $\mathcal{H}_k$. It is defined by

$$\chi_k(q) = \text{tr}\left(q^{L_0 - \frac{c}{24}}\right) \tag{13.7}$$

with $c = 1$ in the case we are investigating. The trace extends over the entire
representation space $\mathcal{H}_k$. For the u(1) current algebra, the character functions can
be computed explicitly, and the result is (see also Chapter 11)

$$\chi_k(q) = q^{\frac{k^2}{2} - \frac{1}{24}}\left(1 + q + 2q^2 + \ldots\right) = q^{\frac{k^2}{2} - \frac{1}{24}} \prod_{n=1}^{\infty} \frac{1}{1 - q^n} =: \frac{q^{\frac{k^2}{2}}}{\eta(q)}, \tag{13.8}$$

where η denotes the DEDEKIND η function we have introduced before. The characters are functions of the variable $q = \exp 2\pi i \tau$, and we allow τ to assume any
value in the upper half $Im\tau \geq 0$ of the complex plane. The set of characters has
a remarkable property: It is invariant under two transformations S, T, which are
known as modular transformations. These act on the variable q according to

$$q \overset{S}{\longmapsto} \tilde{q} = e^{-\frac{2\pi i}{\tau}}, \quad q \overset{T}{\longmapsto} q' = e^{2\pi i(\tau + 1)}. \tag{13.9}$$

Note that our characters involve powers q^κ with some real exponent κ. Hence, we need to distinguish between q' and q. In order to show that the set of characters closes under modular transformations, we need the following two facts about the DEDEKIND η function:

$$\eta(\tilde{q}) = \sqrt{-i\tau}\,\eta(q), \quad \eta(q') = e^{\frac{2\pi i}{24}}\eta(q). \tag{13.10}$$

Using these identities it is then easy to show

$$\chi_k(\tilde{q}) = \frac{e^{-\frac{2\pi i}{\tau}\frac{k^2}{2}}}{\sqrt{-i\tau}\,\eta(q)} = \frac{1}{\eta(q)}\int_{-\infty}^{\infty} dl\, e^{2\pi ikl}e^{2\pi i\tau\frac{l^2}{2}} = \int_{-\infty}^{\infty} dl\, e^{2\pi ilk}\chi_l(q)$$
$$= \int_{-\infty}^{\infty} dl\, S_{kl}\chi_l(q), \tag{13.11}$$

where

$$S_{kl} = e^{2\pi ikl}$$

are elements of the so-called modular S matrix. The second equality is obtained by a simple Gaussian integral. Similarly, we find

$$\chi_k(q') = e^{2\pi i\left(\frac{k^2}{2}-\frac{1}{24}\right)}\chi_k(q) =: \int_{-\infty}^{\infty} dl\, T_{kl}\chi_l(q) \tag{13.12}$$

with a modular T matrix whose matrix elements are

$$T_{kl} = \delta(k-l)e^{2\pi i\left(\frac{k^2}{2}-\frac{1}{24}\right)}.$$

Much of what we have just outlined for the u(1) current algebra remains true for other, more complicated symmetry algebras in 2-dimensional conformal field theories (see below). In particular, the set of character functions for irreducible highest weight representations is known to close under modular transformation, at least for a very wide class of symmetry algebras. This completes our discussion of the representation theory.

13.3 Vertex Operators as Tensor Operators

Our aim is to construct correlation functions of a field theory with u(1) current algebra symmetry. In Chapter 3, the vertex operators V_k were constructed as exponentials of the field X. Hence, the construction of these fields needed the zero mode operator x. Since we do not have x at our disposal (not part of the symmetry), we can no longer proceed in the same way. Instead, we shall now think of V_k as a tensor operator of the u(1) current algebra symmetry; i.e., we shall characterize V_k through its commutation relations with the generators a_n and L_n. As before, we specify these through the commutation relations

$$\left[J_{\gtrless}(z), V_k(w)\right] = \pm \frac{k}{z-w} V_k(w),$$ (13.13)

$$\left[T_{\gtrless}(z), V_k(w)\right] = \pm \left[\frac{h}{(z-w)^2} + \frac{1}{z-w}\partial_w\right] V_k(w)$$ (13.14)

with the semi-infinite sums

$$W_>(z) = W(z) - W_<(z) = \sum_{n>-h} W_n z^{-n-h},$$ (13.15)

which involve all the annihilation operators a_n in the current $W = J$ and L_n in the VIRASORO field $W = T$. Recall that for these two fields, the weights h_W are given by $h_J = 1, h_T = 2$. To a given set of N vertex operators, we would like to assign a function Ω as

$$\Omega = \langle \prod_{\nu=1}^{N} V_{k_\nu}(w_\nu) \rangle.$$ (13.16)

This function then satisfies a set of so-called WARD identities:

$$\langle J(z) \prod_{\nu=1}^{N} V_{k_\nu}(w_\nu) \rangle = \sum_{\nu=1}^{N} \frac{k_\nu}{z-w_\nu} \Omega,$$

$$\langle :J^2:(z) \prod_{\nu=1}^{N} V_{k_\nu}(w_\nu) \rangle = \sum_{\nu,\mu=1}^{N} \frac{k_\nu k_\mu}{(z-w_\nu)(z-w_\mu)} \Omega,$$

$$\langle T(z) \prod_{\nu=1}^{N} V_{k_\nu}(w_\nu) \rangle = \sum_{\nu=1}^{N} \left[\frac{h_\nu}{(z-w_\nu)^2} + \frac{1}{z-w_\nu}\partial_\nu\right] \Omega.$$ (13.17)

They are easily derived by inserting the commutation relations (13.13) and (13.14) after splitting the fields J and T into creation and annihilation operators. The formula $T = \frac{1}{2} : J^2 :$ for the stress energy tensor relates the last two equations in the above list and thereby gives two important equations. First, from the identification of the second order poles, we obtain the conformal dimensions of the vertex operators:

$$h_\nu = \frac{k_\nu^2}{2}.$$ (13.18)

Second, equating the residues of the first order poles, we deduce a simple version of the so-called Knizhnik-Zamolodchikov equations:

$$\partial_\nu \Omega \;=\; \sum_{\mu \neq \nu = 1}^{N} \frac{k_\mu k_\nu}{w_\mu - w_\nu} \Omega. \tag{13.19}$$

Thereby, the symmetry properties of our operators V_k have left us with a set of differential equations for the functions Ω. These equations are easily solved by

$$\Omega \;=\; \text{const} \prod_{\mu \neq \nu = 1}^{N} (w_\mu - w_\nu)^{k_\mu k_\nu} \delta \left(\sum_{\mu = 1}^{N} k_\mu \right). \tag{13.20}$$

The presence of the δ function can be deduced by inserting the zero mode a_0 into our correlation functions; i.e., it is again a consequence of the commutation relations between V_k and a generator of the symmetry algebra. Let us now recall that the theory contains a second u(1) current algebra that is generated by the modes $\bar{a}_n$. Using this symmetry, one can repeat the entire analysis for the anti-holomorphic part of the correlation function. The final outcome for a product of objects $V_{k,\bar{k}} \sim V_k \bar{V}_{\bar{k}}$ is given by

$$\langle \prod_{\nu = 1}^{N} V_{k_\nu, \bar{k}_\nu} (z_\nu, \bar{z}_\nu) \rangle$$

$$= \text{const} \prod_{\mu \neq \nu = 1}^{N} (z_\mu - z_\nu)^{k_\mu k_\nu} (\bar{z}_\mu - \bar{z}_\nu)^{\bar{k}_\mu \bar{k}_\nu} \delta \left(\sum_{\mu = 1}^{N} k_\mu \right) \delta \left(\sum_{\mu = 1}^{N} \bar{k}_\mu \right). \tag{13.21}$$

For the time being, we have not made any assumption about the parameters k and $\bar{k}$ that determine the commutation relations of the vertex operators $V_{k,\bar{k}}$ with the currents J and $\bar{J}$. All functions Ω displayed above are consistent with the two commuting u(1) current algebra symmetries. But they are not all correlation functions of a local quantum field theory. For the latter to be the case, Ω must be single valued on the configuration space $\mathbb{C}^{\otimes N}$. This means in particular that they should not change if one of the arguments $(z_\mu, \bar{z}_\mu)$ is transported around another at $(z_\nu, \bar{z}_\nu)$ and back to its starting point. Therefore, requiring locality leads to

$$e^{2\pi i(k_\mu k_\nu - \bar{k}_\mu \bar{k}_\nu)} \;=\; 1 \quad \Leftrightarrow \quad k_\mu k_\nu - \bar{k}_\mu \bar{k}_\nu \in \mathbb{Z} \tag{13.22}$$

for all pairs $(k_\nu, \bar{k}_\nu)$ and $(k_\mu, \bar{k}_\mu)$. There exists a one-parameter family of solutions for the above conditions. It is parametrized by a real number R. In fact, if all the labels $(k, \bar{k})$ are of the form

$$k \;=\; \frac{n}{R} + \frac{R}{2} w, \quad \bar{k} \;=\; \frac{n}{R} - \frac{R}{2} w \qquad \text{for } n, w \in \mathbb{Z} \tag{13.23}$$

with R being the same for all the N fields, then the condition (13.22) is satisfied. Different fields of the theory are then labeled by different integers n and w. In the limit $R \to \infty$, only operators with $w = 0$ remain well defined, i.e. they possess finite $(k, \bar{k})$. We obtain a theory with $k = \bar{k} \in \mathbb{Z}/R$ whose spectrum approaches the one of the free bosonic field we worked with in the first part of this book. Here, we found that there exists a one-parameter family of local quantum field theories with a u(1) current algebra symmetry. The parameter R has a simple geometric interpretation that we want to turn to next.

13.4 The Compactified Free Boson

The theories that we have just constructed are used to model bosonic strings that move on a circle of radius R. In order to back this interpretation, we recall that the mass spectrum of bosonic string theory is closely related to the operators L_0 and $\bar{L}_0$. Hence, we can learn something about the interpretation of the model by looking at the eigenvalues Δ of $L_0 + \bar{L}_0$.[1] According to eq. (13.18), the states that are created by the vertex operators $V_{k,\bar{k}}$ possess eigenvalues

$$\Delta_{k\bar{k}} := h_k + h_{\bar{k}} = \underbrace{\frac{n^2}{R^2}}_{\text{kinetic energy}} + \underbrace{\frac{R^2}{4} w^2}_{\text{energy of winding}} . \qquad (13.24)$$

There are two series of excitations, which are labeled by n and w. As we see from the form of Δ, the quantity n/R should be interpreted as momentum of a string that moves on a circle of radius R. In fact, the wave function of a plane wave with momentum p on a circle of radius R must be periodic:

$$e^{ipx} \stackrel{!}{=} e^{ip(x+2\pi R)}. \qquad (13.25)$$

This requires p to be of the form $p = n/R$ with some integer n. A string that winds around the circle of radius R has an energy that is proportional to the square of its length wR, where w counts the number of times the string wraps around the circle. Both the kinetic and winding energy of a string on a circle of radius R are present in the expression (13.24).

We want to add one important observation concerning the set of eigenvalues (13.24) of $L_0 + \bar{L}_0$ in our model. As the formula shows, the spectrum is actually the same in the models with radius R and $2/R$. In order to make the formulas for eigenvalues match, we only have to exchange the labels n and w. This symmetry of

---

[1] Note that we do not impose the physical state condition $L_0 = \bar{L}_0$ because in a full string theory, both L_0 and $\bar{L}_0$ receive extra terms from other directions in space-time. The selection of physical states can only be performed within the full theory.

the spectrum, which is widely known as T-duality, has a simple explanation. For a large radius R of our circle, the momentum modes are narrowly spaced, whereas we have wide gaps between winding modes. As we bring the radius down, the energy gap between winding modes comes down while momentum modes become heavier. At some point – the radius $R = \sqrt{2}$ – both types of modes possess the same gaps. If we go beyond this point, we observe one set of gaps decreasing while the other increases. The labeling by momentum and winding number is a matter of choice.

The self-dual radius $R_{sd} = \sqrt{2}$ is very special. The reader is left with the exercise of showing that this theory contains two additional fields $J^{\pm} = V_{\pm\sqrt{2},0}$ with conformal dimensions $(h, \bar{h}) = (1, 0)$. These fields fulfill

$$J^{+}(z)J^{-}(w) = \frac{1}{(z-w)^2} + \frac{1}{z-w}J(w) + \mathcal{O}((z-w)^0). \tag{13.26}$$

Together with the u(1) current $J(z)$ the fields $J^{\pm}(z)$ generate what is known as a su(2) current algebra. We shall see this algebra again in one of the following chapters.

Exercises

Problem 28. *Consider the theory of a single free boson at compactification radius $R_{sd} = \sqrt{2}$.*

(a) Demonstrate that this theory contains exactly two vertex operators $V_{k\bar{k}}$ of conformal weight $(h, \bar{h}) = (1, 0)$. As in the text above, we shall denote these by $J^{\pm}$.

(b) Verify the following two relations:

$$[J_{>}(z), J^{\pm}(w)] = \pm\frac{\sqrt{2}}{z-w}J^{\pm}(w), \tag{13.27}$$

$$J^{+}(z)J^{-}(w) \sim \frac{1}{(z-w)^2} + \frac{\sqrt{2}}{z-w}J(w) + \cdots, \tag{13.28}$$

where we omitted all regular terms.
Hint: In order to show the second relation, one should insert the product $J^{+}(z)J^{-}(w)$ into a correlation function with other vertex operators and expand in the difference $z - w$, omitting all the regular terms.

Problem 29. *Consider a pair $\Psi = (\psi_1, \psi_2)$ of real fermions in the NS-sector and define the following three fields:*

$$J^{a}(z) =: \Psi(z)\sigma^{a}\Psi^{t}(z):$$

where σ^a are the Pauli matrices:

$$\sigma^+ = \begin{pmatrix} 0 & 1 \\ 0 & 0 \end{pmatrix}, \quad \sigma^- = \begin{pmatrix} 0 & 0 \\ 1 & 0 \end{pmatrix}, \quad \sigma^0 = \begin{pmatrix} 1/2 & 0 \\ 0 & -1/2 \end{pmatrix}.$$

Determine the commutator $[J_n^+, J_m^-]$.

HINT: Distinguish between the cases $n + m \neq 0$ and $n + m = 0$ and make sure your expressions are normal ordered in the latter case.

14

Modular Invariants and Orbifolds

In the last chapter we have found a family of local quantum fields theories with u(1) current algebra symmetry. The analysis we had to go through, however, was quite involved. In fact, we had to use the symmetry properties of vertex operators to derive differential equations for correlation functions. These then needed to be solved before we were able to impose the locality condition. The aim of the present chapter is to replace the locality condition for correlation functions on the complex plane by the condition of modular invariance of the so-called partition function. Modular invariant partition functions are comparatively easy to find. As an example we shall construct the spectrum of orbifold conformal field theories. These are used to model strings in quotient spaces X/G, where G is some discrete group of finite order that acts on the target space X.

14.1 Partition Function and Modular Invariance

Given any 2-dimensional conformal field theory, such as the models we constructed in the previous chapter, one may compute the so-called partition function. It is given by

$$\mathcal{Z}(q, \bar{q}) \equiv \mathcal{Z}(\tau) = tr_{\mathscr{H}} \left(e^{-2\pi \tau_2 H + 2\pi i \tau_1 P} \right) \tag{14.1}$$

$$= tr_{\mathscr{H}} \left(q^{L_0 - \frac{c}{24}} \bar{q}^{\bar{L}_0 - \frac{c}{24}} \right) \tag{14.2}$$

The quantity is defined for a complex variable $\tau = \tau_1 + i\tau_2$ in the upper half plane, i.e., with $\tau_2 = Im\tau \geq 0$. The latter condition must be imposed for the sum over states to converge, and it assumes that the spectrum of the Hamiltonian $H = L_0 + \bar{L}_0 - c/12$ is bounded from below. The operator $P = L_0 - \bar{L}_0$ is the generator of translations of the world-sheet variable σ. In the second line we have rewritten the first line in terms of L_0 and $\bar{L}_0$, and we traded the parameter τ for $q = \exp(2\pi i\tau)$. For $\tau_1 = 0$, the quantity $\mathcal{Z}$ is the usual finite temperature partition function with temperature $T = 1/\beta = 1/2\pi\tau_2$. For $\tau_1 \neq 0$, the argument of the trace is accompanied by a translation in space. In a local theory on the circle that is parametrized by $\sigma \in [0, 2\pi]$, a translation by 2π acts trivially, i.e., $\exp(2\pi i P) = 1$. Hence, we conclude that

$$(T\mathcal{Z})(\tau) := \mathcal{Z}(\tau + 1) = \mathcal{Z}(\tau). \tag{14.3}$$

In a 2-dimensional conformal field theory, the quantity $\mathcal{Z}(\tau)$ has another important symmetry property. namely it is also invariant under the modular S transformation:

$$(S\mathcal{Z})(\tau) := \mathcal{Z}\left(-\frac{1}{\tau}\right) = \mathcal{Z}(\tau). \tag{14.4}$$

We will not argue for this invariance in general, but check that it is indeed obeyed by the partition function $\mathcal{Z}_R$ of the compactified free boson:

$$\mathcal{Z}_R(q, \bar{q}) = tr_{\mathcal{H}}\left(q^{L_0 - \frac{c}{24}} \bar{q}^{\bar{L}_0 - \frac{c}{24}}\right) = \sum_{n,w\in\mathbb{Z}} \chi_{\frac{n}{R} + \frac{R}{2}w}(q) \chi_{\frac{n}{R} - \frac{R}{2}w}(\bar{q}). \tag{14.5}$$

Here we have used the decomposition $\mathcal{H} = \bigoplus_{n,w} \mathcal{H}_k \otimes \overline{\mathcal{H}}_{\bar{k}}$ of the state space into representations of the left- and right-moving u(1) current algebras. The dependence of the labels $k, \bar{k}$ on n and w is given through eq. (13.23), and we defined the characters χ_k in eq. (13.8). Under the modular S transformation the partition functions $\mathcal{Z}_R$ behaves as

$$(S\mathcal{Z}_R)(q, \bar{q}) = \mathcal{Z}_R(\tilde{q}, \bar{\tilde{q}}) = \sum_{n,w\in\mathbb{Z}} \chi_{\frac{n}{R} + \frac{R}{2}w}(\tilde{q}) \, \chi_{\frac{n}{R} - \frac{R}{2}w}(\bar{\tilde{q}})$$

$$= \int_{\mathbb{R}^2} dl \, d\bar{l} \sum_{n,w\in\mathbb{Z}} e^{2\pi i n \frac{l - \bar{l}}{R}} e^{2\pi i w \frac{R}{2}(l + \bar{l})} \chi_l(q) \, \chi_{\bar{l}}(\bar{q})$$

$$= \int_{\mathbb{R}^2} dl \, d\bar{l} \, 2 \sum_{\tilde{n}, \tilde{w}\in\mathbb{Z}} \delta\left(l + \bar{l} - \frac{2}{R}\tilde{n}\right) \delta(l - \bar{l} - R\tilde{w}) \, \chi_l(q) \, \chi_{\bar{l}}(\bar{q})$$

$$= \sum_{\tilde{n}, \tilde{w}\in\mathbb{Z}} \chi_{\frac{\tilde{n}}{R} + \frac{R}{2}\tilde{w}}(q) \, \chi_{\frac{\tilde{n}}{R} - \frac{R}{2}\tilde{w}}(\bar{q}) = \mathcal{Z}_R(q, \bar{q}).$$

In passing to the second line, we have used the behavior (13.11) of the characters $\chi_k(q)$ under modular S transformation. Then, we carried out the sum over n and w using a special case of the so-called POISSON resummation formula:

$$\sum_{n\in\mathbb{Z}} e^{2\pi i a x n} = \sum_{\tilde{n}\in\mathbb{Z}} \frac{1}{|a|} \delta(x - \tilde{n}/a). \tag{14.6}$$

In the last line, we finally performed the integral over l and $\bar{l}$. Hence, we conclude that the partition function of a compactified free boson is invariant under modular S and T transformations.

Looking at our short computation, we can trace the modular invariance back to the locality condition $kl - \bar{k}\bar{l} \in \mathbb{Z}$ that we solved through the Ansatz (13.23). In other words, the condition (13.22) that we derived by imposing locality of correlation functions can also be obtained from modular invariance of the associated partition function. Since the analysis of modular invariance was significantly easier than the study of locality, it is tempting to detect consistent conformal field theories through the existence of a modular invariant partition function. From now on we shall indeed adopt modular invariance as a fundamental principle and trust that it selects consistent models.

In order to see the principle of modular invariance at work and build some more confidence in it, we want to briefly discuss the space of parameters for a model that involves D compactified free bosons. We write the partition function as

$$\mathcal{Z}_\Xi(\tau) = \sum_{(k,\bar{k}) \in \Xi} \chi_k(q) \, \chi_{\bar{k}}(\bar{q}). \tag{14.7}$$

Here, $k \in \mathbb{R}^D$ and $\bar{k} \in \mathbb{R}^D$ are D-dimensional momenta for left- and right-moving modes, respectively, and the set $\Xi \subset \mathbb{R}^{D,D}$ must be chosen such that $\mathcal{Z}_\Xi$ is modular invariant. We have introduced $\chi_k = \prod_\nu \chi_{k_\nu}(q)$ to denote a product of characters of the u(1) current algebra. It is not difficult to evaluate the conditions on the set Ξ that follow from the invariance of $\mathcal{Z}$ under modular T and S tranformations,

$$k^2 - \bar{k}^2 \in 2\mathbb{Z} \ , \quad kl - \bar{k}\bar{l} \in \mathbb{Z}, \tag{14.8}$$

for all pairs $(k,\bar{k}), (l,\bar{l}) \in \Xi$. The allowed vectors $(k,\bar{k}) \in \Xi$ form a $2D$-dimensional lattice. Suppose we are given one such lattice Ξ_*. Then it is easy to construct a whole family $\Xi_H = H\Xi_*$ by acting with an element $H \in SO(D,D)$. Indeed, the conditions (14.8) are preserved by the action of $SO(D,D)$. The transformed lattice Ξ_H does not always give rise to a new theory. In fact, if two lattices Ξ_H and $\Xi_{H'}$ are related by a transformation in $SO(D) \times SO(D) \subset SO(D,D)$, then the corresponding partition functions are identical. Therefore, consistent conformal field theories on a D-dimensional torus are parametrized by the coset space:

$$\mathcal{M} = \frac{SO(D,D)}{SO(D) \times SO(D)}. \tag{14.9}$$

The dimension of $\mathcal{M}$ is given by *dim* $\mathcal{M} = (2D - 1)D - 2(D - 1)D/2 = D^2$. One may wonder about the meaning of all these parameters. If all we could change were the radii of the D circles, we would expect D parameters rather than D^2. The explanation of our findings is not difficult. Let us denote the D bosonic fields by

ϕ_ν, $\nu = 1, \ldots, D$. These fields take values in the circle, i.e., $\phi_\nu(z, \bar{z}) \in [0, 2\pi]$. The most general action that is consistent with the current algebra symmetry takes the form

$$S[\phi_\nu] = \frac{1}{2\pi\alpha'} \int d^2z \, \Omega^{\mu\nu} \, \partial\phi_\nu \bar{\partial}\phi_\mu$$

with a constant $D \times D$ matrix Ω. The matrix elements of Ω contain the D^2 parameters we discovered through our search for possible modular invariant partition functions.

14.2 The Modular Group and Complex Tori

The transformation S and T were introduced in the previous section through their action on the parameter τ in the upper half

$$\mathbb{H} = \{ \tau \in \mathbb{C} \mid Im\tau \geq 0 \} \tag{14.10}$$

of the complex plane. Repeated action of S and T generates a group of symmetry transformations of $\mathbb{H}$. This group is called the *modular group* and is given by

$$\text{PSL}(2, \mathbb{Z}) = \left\{ \gamma = \begin{pmatrix} a & b \\ c & d \end{pmatrix} \middle| a, b, c, d \in \mathbb{Z}; ad - bc = 1 \right\} \middle/ \{\gamma \sim -\gamma\}.$$

The group PSL(2,$\mathbb{Z}$) of such matrices acts on the upper half of the complex plane through the usual rational transformation:

$$\gamma(\tau) = \frac{a\tau + b}{c\tau + d}. \tag{14.11}$$

It is easy to see that the transformations are consistent with the group multiplication in PSL(2,$\mathbb{Z}$). The modular transformations T and S are special elements of the modular group. They are given by

$$T = \begin{pmatrix} 1 & 1 \\ 0 & 1 \end{pmatrix} \sim \begin{pmatrix} -1 & -1 \\ 0 & -1 \end{pmatrix}, \quad S = \begin{pmatrix} 0 & 1 \\ -1 & 0 \end{pmatrix} \sim \begin{pmatrix} 0 & -1 \\ 1 & 0 \end{pmatrix}. \tag{14.12}$$

It requires very little work to show that the transformations T, S generate all of PSL(2,$\mathbb{Z}$).

From the modular invariance of the partition functions $\mathcal{Z}(\tau)$ under S and T transformations we can conclude that $\mathcal{Z}(\tau)$ is invariant under all modular transformations $\gamma \in \text{PSL}(2,\mathbb{Z})$:

$$\mathcal{Z}(\gamma(\tau)) = \mathcal{Z}(\tau).$$

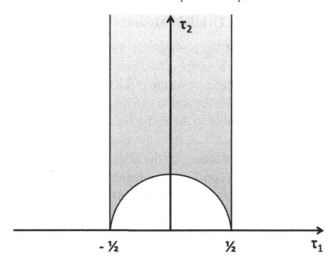

Figure 14.1 The fundamental domain.

The property implies that $\mathcal{Z}$ may be considered as a function on the quotient space $\mathbb{H}/\mathrm{PSL}(2,\mathbb{Z})$. One may picture this quotient space by selecting one representative of each $\mathrm{PSL}(2,\mathbb{Z})$ orbit in the upper half-plane. The most common choice is shown in Figure 14.1.

Before we move on we want to show that the quotient space parametrizes inequivalent complex tori. In order to construct a complex torus, we depart from the complex plane $\mathbb{C}$ and consider the action of translations by elements of the lattice

$$\Gamma_\tau = \mathbb{Z} + \tau\mathbb{Z} = \{ z \in \mathbb{C} \,|\, z = n + \tau m; \; n, m \in \mathbb{Z} \}, \qquad (14.13)$$

where $\tau \in \mathbb{H}$ is taken to be any complex number in the upper half plane. All complex tori are of the form $T_\tau = \mathbb{C}/\Gamma_\tau$ for some choice of τ. Closer inspection reveals that different choices of τ may actually lead to the same complex torus. Suppose, for example, that we switch from τ to $\tau + 1$. Such a substitution leaves the corresponding lattice Γ_τ unaltered, $\Gamma_\tau = \Gamma_{\tau+1}$. Hence, it has no effect on the torus $\mathbb{C}/\Gamma_\tau$. More generally, two lattices Γ_τ and $\Gamma_{\tau'}$ are identical[1] if and only if the parameters τ and τ' are related by a modular transformation, i.e., iff $\tau' = \gamma(\tau)$ for some $\gamma \in \mathrm{PSL}(2,\mathbb{Z})$. Putting all this together we have now established that the counting function $Z(\tau)$ of our local field theory is a function on the moduli space of complex tori.

---

[1] Identical means that they can be mapped to each other by a holomorphic map of the complex plane, in the case at hand by a simple multiplication with $c\tau + d$.

14.3 The Orbifold Modular Invariant

As an important application of the principle of modular invariance we now want
to describe the so-called orbifold construction that allows us to build new modular
invariant partition functions whenever we are given the state space $\mathscr{H}$ of some local
conformal field theory along with the action of a finite group G on $\mathscr{H}$; see [19] for a
detailed discussion and references to the early literature. If $\mathscr{H}$ describes the states
of a field theory with target space X, then the modular invariant we are about to
construct counts states of a field theory on the orbifold X/G. The definition of X/G
requires fixing some action of G on X. Let us denote the image of some point $x \in X$
under the action of $g \in G$ by $gx \in X$. Then we can define

$$X/G = \{ Gx \mid x \in X \}.$$

Here Gx denotes the orbit of $x \in X$ under the action of G. As an example, we may
consider $X = \mathbb{R}$ and $G = \mathbb{Z}_2$ such that the non-trivial element $g \in G$ acts by a
reflection at the origin, $gx = -x$. In this case, the quotient space $X/G = \mathbb{R}/\mathbb{Z}_2 =
\mathbb{R}_0^+$ is given by the half-line. As we can see from this simple example, the quotient
of a smooth manifold need not be smooth again.

In order to proceed, let us provide a slightly more precise formulation of the
data we begin with. Suppose we are given the state space $\mathscr{H}$ of a conformal field
theory. As usual, the space $\mathscr{H}$ carries a representation of the VIRASORO algebra
with a central charge c. We suppose that this state space is also acted upon by some
discrete group G, i.e. that there exists a representation π,

$$\pi : G \rightarrow End(\mathscr{H})$$

$$g \mapsto \big(\pi(g) : \mathscr{H} \rightarrow \mathscr{H}\big), \tag{14.14}$$

such that $\pi(g)$ commutes with the VIRASORO modes L_n and $\bar{L}_n$ for all $n \in \mathbb{Z}$. Put
differently, this means that the action with elements from G leaves the stress energy
tensor invariant.

Given these data we now want to introduce some modular invariant partition
function. If $\mathscr{H}$ was the state space of some G-symmetric quantum mechanical
system, we could easily find those states of the theory that respect the action of
G. All we would have to do was to select those elements of $\mathscr{H}$ that are invariant
under the action of G. There exists a standard operator that performs this projection
for us:

$$P = \frac{1}{|G|} \sum_{g \in G} \pi(g). \tag{14.15}$$

Here, $|G|$ denotes the number of elements in G. It is easy to see that P is a projection, $P^2 = P$, onto G invariant states. Consequently, G-invariant states in $\mathcal{H}$ are counted by the following function:

$$\mathcal{Z}^{\text{proj}}(\tau) = \text{tr}_{\mathcal{H}}\left(Pq^{L_0-\frac{c}{24}}\bar{q}^{\bar{L}_0-\frac{c}{24}}\right) = \frac{1}{|G|}\sum_{g\in G}\text{tr}_{\mathcal{H}}\left(\pi(g)q^{L_0-\frac{c}{24}}\bar{q}^{\bar{L}_0-\frac{c}{24}}\right).$$

(14.16)

This auxiliary function $\mathcal{Z}^{\text{proj}}$ counts the number of G-invariant states in the state space $\mathcal{H}$ of the conformal field theory. But $\mathcal{Z}^{\text{proj}}$ is not the partition function of some new conformal field theory. As we will see in some concrete examples below, $\mathcal{Z}^{\text{proj}}$ fails to be modular invariant. Nevertheless, it is a useful starting point to construct such an invariant.

The rough idea is to simply take $\mathcal{Z}^{\text{proj}}(\tau)$ and to average it over the modular group:

$$\mathcal{Z}^{\text{orb}}(\tau) \ :\sim \ \sum_{\gamma\in\text{PSL}(2,\mathbb{Z})}\mathcal{Z}^{\text{proj}}(\gamma(\tau)).$$

By construction, $\mathcal{Z}^{\text{orb}}$ is a modular invariant. The main issue with the previous formula is to actually make it well defined. Note that the modular group is infinite so that the sum of γ is infinite as well. As long as G is finite, the sum over γ contains a finite number of different contributions only, each with infinite multiplicity. This multiplicity must be removed by hand such that the coefficient of $|q|^{-\frac{c}{12}}$ is unity.

We can actually give a somewhat more precise formula that avoids such issues. It turns out that the orbifold partition function $\mathcal{Z}^{\text{orb}}$ may be written as

$$\mathcal{Z}^{\text{orb}}(\tau) = \frac{1}{|G|}\sum_{(g,h)_{cp}}\mathcal{Z}_{g,h}(\tau),$$

(14.17)

where the sum extends over all commuting pairs $(g,h)_{cp} \in \{(g,h) \in G \times G \,|\, gh = hg\}$. The functions $\mathcal{Z}_{g,h}$ are uniquely characterized by the following properties:

$$\mathcal{Z}_{g,e}(\tau) = \text{tr}_{\mathcal{H}}\left(\pi(g)q^{L_0-\frac{c}{24}}\bar{q}^{\bar{L}_0-\frac{c}{24}}\right),$$

(14.18)

where e denotes the unit element of the orbifold group and

$$T\mathcal{Z}_{g,h}(\tau) = \mathcal{Z}_{g,h}(\tau+1) = \mathcal{Z}_{gh,h}(\tau),$$

(14.19)

$$S\mathcal{Z}_{g,h}(\tau) = \mathcal{Z}_{g,h}(-1/\tau) = \mathcal{Z}_{h,g}(\tau).$$

(14.20)

We will show below how these formulas can be put to work in a concrete example. One simply uses the definition in eq. (14.18) to construct $\mathcal{Z}_{g,e}$ and then finds the remaining functions for non-trivial commuting pairs (g, h) through repeated application of the formulas (14.19) and (14.20). With the help of these formulas one may also show that the function on the right-hand side of eq. (14.17) is modular invariant. Note that we can split off our original counting function $\mathcal{Z}^{\text{proj}}$ such that

$$\mathcal{Z}^{\text{orb}}(\tau) \;=\; \mathcal{Z}^{\text{proj}}(\tau) \;+\; \sum_{(g,h)_{cp}; h \neq e} \mathcal{Z}_{g,h}(\tau).$$

Therefore, the modular invariant $\mathcal{Z}^{\text{proj}}$ may be considered as a counting function for a state space that contains more than just those states from $\mathcal{H}$ that are invariant under the action of our orbifold group G.

One may wonder whether one can actually construct the state space of the orbifold theory rather than its counting function $\mathcal{Z}^{\text{orb}}$. This is indeed possible. The basic idea is to quantize our original field theory with h-twisted boundary conditions. One way of expressing what we mean by twisted boundary conditions makes use of the state field correspondence. Recall that we can actually associate a field $V_\psi(\tau, \sigma)$ to every state $\psi \in \mathcal{H}$. So far, these fields always satisfied periodic boundary conditions $V(\sigma + 2\pi) = V(\sigma)$. In the case of of h-twisted boundary conditions we demand that

$$V_\psi(\sigma + 2\pi) \;=\; V_{\pi(h)\psi}(\sigma) \quad \text{for all} \quad \psi \in \mathcal{H}. \tag{14.21}$$

Here, the field $\pi(h)\psi$ denotes the action of $h \in G$ on the state ψ. The space that is generated from the vacuum by application of such fields with h-twisted boundary conditions will be denoted by $\mathcal{H}_h$. After these preparations, we can now construct the functions $\mathcal{Z}_{(g,h)}$ through

$$\mathcal{Z}_{g,h}(\tau) \;=\; \text{tr}_{\mathcal{H}_h}\left(\pi(g) q^{L_0 - \frac{c}{24}} \bar{q}^{\bar{L}_0 - \frac{c}{24}}\right). \tag{14.22}$$

Note that this formula generalizes our definition of the functions $\mathcal{Z}_{g,e}$ in eq. (14.18).

14.4 Example: The Orbifold $\mathbb{R}/\mathbb{Z}_2$

To illustrate many of the statements and constructions we have described in the previous section, we shall now discuss the orbifold partition function for the quotient $\mathbb{R}/\mathbb{Z}_2 \cong \mathbb{R}_0^+$:

We begin by spelling out the basic input into the construction: the state space $\mathcal{H}$ and the action of $\mathbb{Z}_2$ on it. Our state space is the usual space we have worked with in our treatment of bosonic string theory:

$$\mathcal{H} = \int_{-\infty}^{\infty} dk \, \mathcal{H}_k \otimes \overline{\mathcal{H}}_k. \tag{14.23}$$

The spaces $\mathcal{H}_k$ and $\overline{\mathcal{H}}_k$ are constructed out of left- and right-moving ground states $|k\rangle$ and $|\bar{k}\rangle$ by application of the creation operators a_{-n} and $\bar{a}_{-n}$, respectively. On this space we have constructed the bosonic field $X(z, \bar{z})$ in Chapter 3:

$$X(z, \bar{z}) = x - \frac{i}{2}\alpha' p \ln(z\bar{z}) + i\sqrt{\frac{\alpha'}{2}} \sum_{n \neq 0} \left(\frac{a_n}{n} z^{-n} + \frac{\bar{a}_n}{n} \bar{z}^{-n} \right). \tag{14.24}$$

Now we would like to introduce an action $\pi(g)$ of the non-trivial element $g \in \mathbb{Z}_2 = \{e, g\}$ on $\mathcal{H}$ such that $\pi(g)X(z, \bar{z}) = -X(z, \bar{z})\pi(g)$. The condition implies that

$$\pi(g)\,|k\rangle \otimes |\bar{k}\rangle = |-k\rangle \otimes |-\bar{k}\rangle, \quad \pi(g)\,a_n = -a_n\pi(g), \quad \pi(g)\,\bar{a}_n = -\bar{a}_n\pi(g). \tag{14.25}$$

The action of $\pi(g)$ on the ground states along with the commutation relations between $\pi(g)$ and the creation operators determine the action of $\pi(g)$ on the entire state space.

Having fixed our basic data we want to determine the counting functions $\mathcal{Z}^{\text{proj}}$ for $\mathbb{Z}_2$-invariant states in $\mathcal{H}$:

$$\mathcal{Z}^{\text{proj}}(\tau) = \frac{1}{2} \text{tr} \left((1 + \pi(g)) q^{L_0 - \frac{1}{24}} \bar{q}^{\bar{L}_0 - \frac{1}{24}} \right)$$

$$= \frac{1}{2} \left(\mathcal{Z}^{\mathbb{R}}(\tau) + \left| q^{-\frac{1}{24}} \prod_{n=1}^{\infty} \frac{1}{1 + q^n} \right|^2 \right) = \frac{1}{2}\mathcal{Z}^{\mathbb{R}}(\tau) + \left| \frac{\eta(\tau)}{\theta_2(\tau)} \right|. \tag{14.26}$$

The result requires a bit of explanation. The function $\mathcal{Z}^{\mathbb{R}}$ is the partition function of the free bosonic field theory. We do not need to restate its form here. All we need to know about $\mathcal{Z}^{\mathbb{R}}$ is that it transforms trivially under modular S and T transformation. The second term needs some insight into counting functions for free bosonic fields. Let us first note that

$$\text{tr}_{\mathcal{H}_0 \otimes \overline{\mathcal{H}}_0} \left(q^{L_0 - \frac{1}{24}} \bar{q}^{\bar{L}_0 - \frac{1}{24}} \right) = \left| q^{-\frac{1}{24}} \prod_{n=1}^{\infty} \frac{1}{1 - q^n} \right|^2.$$

In contrast to a similar formula we derived in Chapter 11, we have included a left-moving sector now. This is why the result appears with the absolute square $| \cdot |^2$.

After we insert the operators $\pi(g)$ into the trace, the counting works in the same way, only that we have to change the sign in the denominator:

$$\text{tr}_{\mathcal{H}_0 \otimes \overline{\mathcal{H}}_0}\left(\pi(g)q^{L_0-\frac{1}{24}}\bar{q}^{\bar{L}_0-\frac{1}{24}}\right) = \left| q^{-\frac{1}{24}} \prod_{n=1}^{\infty} \frac{1}{1+q^n} \right|^2.$$

The infinite product $\prod(1+q^n)$ is a close relative of one of the JACOBI theta functions, the function θ_2. The precise relation was given previously in eq. (11.27). From the same chapter we also know two similar relations, eqs. (11.28) and (11.29), that involve products with half-integers powers of q.

In order to proceed, we will need to know how the functions θ_i/η transform under modular S and T transformations. The results can be found in the following table.

	S	T
$\sqrt{\dfrac{\theta_2}{\eta}}$	$\sqrt{\dfrac{\theta_4}{\eta}}$	$e^{\frac{i\pi}{12}}\sqrt{\dfrac{\theta_2}{\eta}}$
$\sqrt{\dfrac{\theta_3}{\eta}}$	$\sqrt{\dfrac{\theta_3}{\eta}}$	$e^{-\frac{i\pi}{24}}\sqrt{\dfrac{\theta_4}{\eta}}$
$\sqrt{\dfrac{\theta_4}{\eta}}$	$\sqrt{\dfrac{\theta_2}{\eta}}$	$e^{-\frac{i\pi}{24}}\sqrt{\dfrac{\theta_3}{\eta}}$

Let us now come back to the issue of modular invariance of the counting function (14.26). From the results in the table we conclude that $\mathcal{Z}^{\text{proj}}$ is invariant under modular T transformation. But it fails to be invariant under modular S transformations, in agreement with our general claims in the previous section. In order to construct a modular invariant we must follow the general recipe we sketched at the end of the previous section.

Our task is to produce a function $\mathcal{Z}_{g,h}(\tau)$ for all commuting pairs of elements (g,h) in $G = \mathbb{Z}_2$. Since the orbifold group is Abelian, we admit all pairs (g,h), i.e., the pairs (e,e), (e,g), (g,e), and (g,g). Two of these counting functions are given by eq. (14.18). In the case at hand, these read

$$\mathcal{Z}_{e,e}(\tau) = \mathcal{Z}^{\text{R}}(\tau), \quad \mathcal{Z}_{g,e}(\tau) = \left| \frac{2\eta(\tau)}{\theta_2(\tau)} \right|. \tag{14.27}$$

The two remaining functions can be obtained by application of the modular S and T transformations. We find that

$$\mathcal{Z}_{e,g}(\tau) = S\mathcal{Z}_{g,e}(\tau) = \left| \frac{2\eta(\tau)}{\theta_4(\tau)} \right|, \quad \mathcal{Z}_{g,g}(\tau) = T\mathcal{Z}_{e,g} = \left| \frac{2\eta(\tau)}{\theta_3(\tau)} \right|. \tag{14.28}$$

The final expression (14.17) for the orbifold partition function of the theory on $\mathbb{R}/\mathbb{Z}_2$ is then

$$\mathcal{Z}^{\text{orb}}(\tau) = \frac{1}{2}\left(\mathcal{Z}^{\mathbb{R}}(\tau) + \left|\frac{2\eta(\tau)}{\theta_2(\tau)}\right| + \left|\frac{2\eta(\tau)}{\theta_3(\tau)}\right| + \left|\frac{2\eta(\tau)}{\theta_4(\tau)}\right|\right). \qquad (14.29)$$

The states that are to be found in $\mathcal{Z}_{e,e}$ and $\mathcal{Z}_{g,e}$ were present in our original theory. Those that are associated with $\mathcal{Z}_{e,g} + \mathcal{Z}_{g,g}$, however, are new. Geometrically, they may be understood as string states in the space $\mathbb{R}$ that have one end at some point x and the other at the reflected point $-x$; see the figure below. These look like closed strings only after the identification between x and $-x$ has been performed:

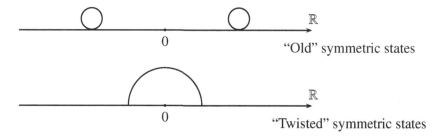

0 "Old" symmetric states

0 "Twisted" symmetric states

In the case at hand we can describe the new states quite explicitly. To this end, we solve the 2-dimensional wave equation $\Delta X = 0$ with antiperiodic boundary conditions. The most general solution satisfying $X(\sigma + 2\pi, \tau) = -X(\sigma, \tau)$ is of the form

$$X(z, \bar{z}) = i\sqrt{\frac{\alpha'}{2}} \sum_{r \in \mathbb{Z} + \frac{1}{2}} \frac{1}{r}\left(a_r z^{-r} + \bar{a}_r \bar{z}^{-r}\right). \qquad (14.30)$$

Following our standard procedure, we quantize this theory by imposing the commutation relations $[a_r, a_s] = ir\delta_{r+s,0}$ for the modes of X. Note that the solution has no zero modes. This is in agreement with the geometric intuition: a string that has one end at x and the other at $-x$ cannot move the position of its center of mass, it is pinned to the origin. The modes a_r and $\bar{a}_r$ may be split into creation and annihilation operators. The former are used to generate the state space $\mathcal{H}_g$ from ground states $|\eta_\pm\rangle$ with properties

$$a_r|\eta_\pm\rangle = 0 = \bar{a}_r|\eta_\pm\rangle \quad \text{for all } r \geq \frac{1}{2}$$

by application of $a_r, \bar{a}_r$ with $r \leq -\frac{1}{2}$. The two ground states are associated to the two different orientations of strings at the fixed point. On the resulting state space we can define the action of two commuting copies of the VIRASORO algebra with central charge $c = 1$,

$$L_n = \frac{1}{2} \sum_{r \in \mathbb{Z} + \frac{1}{2}} : a_{n-r} a_r : + \frac{1}{16} \delta_{n,0}, \qquad (14.31)$$

and similarly for $\bar{L}_n$. The constant $1/16$ must be added to the zero mode L_0 in order to obtain the usual commutation relations of the VIRASORO algebra; see Chapter 8 for a similar discussion in the case of fermions. Using the same kind of reasoning as in our discussion above, it is easy to derive that

$$\frac{1}{2} \operatorname{tr}_{\mathscr{H}_g} \left((1 + \pi(g)) q^{L_0 - \frac{1}{24}} \bar{q}^{\bar{L}_0 - \frac{1}{24}} \right)$$

$$= \left| q^{\frac{1}{48}} \prod_{n=1}^{\infty} \frac{1}{1 - q^{n-1/2}} \right|^2 + \left| q^{\frac{1}{48}} \prod_{n=1}^{\infty} \frac{1}{1 + q^{n-1/2}} \right|^2 = \left| \frac{\eta(\tau)}{\theta_3(\tau)} \right| + \left| \frac{\eta(\tau)}{\theta_4(\tau)} \right|.$$

$$(14.32)$$

We see that the trace over states in the twisted sector $\mathscr{H}_g$ reproduces the two terms in eq. (14.29) that were not included in the counting function (14.26) for $\mathbb{Z}_2$ invariant states in $\mathscr{H}$.

Exercises

Problem 30. *Use the properties* (14.19) *and* (14.20) *to show that the function defined through eq.* (14.17) *is modular invariant.*

Problem 31. *Show that the generators L_n that are defined in eq.* (14.31) *satisfy the defining relations of the Virasoro algebra with central charge $c = 1$.*

15

Closed Strings on the 3-Sphere

The aim of this chapter is to describe at least one non-trivial background that is based on a truly interacting 2-dimensional field theory. The example we picked is one with a constant curvature metric: the 3-sphere S^3. It is arguably the simplest geometric background that gives rise to a non-trivial 2-dimensional field theory with VIRASORO symmetry. The 3-sphere S^3 has many nice features. In particular, it may be identified with the group manifold $S^3 = \mathrm{SU}(2)$. Models in which the basic fields take values on a group manifold have become the paradigmatic examples to illustrate the power of the technology we introduced over the last two chapters and, in a certain sense, all known 2-dimensional conformal field theories can be considered derivatives thereof.

By itself, the 3-sphere cannot be a consistent string background, simply because it is not 26(10)-dimensional. But it (or some close relatives) does appear as building block for superstring compactifications. Through the so-called GEPNER construction, the quotient $\mathrm{SU}(2)/\mathrm{U}(1)$ enters the construction of CALABI-YAU compactifications deep in the stringy regime. Furthermore, 3-spheres are also part of certain AdS compactifications such as $AdS_3 \times S^3 \times T^4$. Finally, the 3-sphere theory we are about to solve is an important component in the near horizon geometry of NS5-branes; see [40].

As in the previous discussion, it is actually useful to gain some first experience by looking at particles on the group manifold SU(2). We will do so in the first section before introducing the 2-dimensional field theory in the second. This theory is then solved in the third section. Obviously, our presentation has to remain a bit sketchy in parts. We refer the reader to the very comprehensive discussion in [18] for more details and many interesting extensions.

15.1 Particle on the Group Manifold SU(2)

Our ultimate goal is to construct the space of states for a field theory in which the basic fields take values on $S^3 \cong$ SU(2). What we are about to build is the modular invariant partition function of the system. Before we get there, it seems worthwhile understanding the spectrum of a particle that moves freely on SU(2). Our presentation here might seem a bit overly formal, but it is designed so as to anticipate some of the constructions we shall later employ when dealing with the full field theory.

Before we can actually start our analysis we need to construct the Hamiltonian of a particle on the 3-sphere. It is the counterpart of the field theoretic Hamiltonian $H \sim L_0 + \bar{L}_0$ that encodes the mass spectrum of string theory. The particle Hamiltonian is related to the LAPLACIAN Δ on S^3 through

$$H = \frac{1}{2m}L^2 = -\frac{\hbar^2}{2m}\Delta. \tag{15.1}$$

The LAPLACIAN is constructed from left- or right-invariant vector fields on the group manifold SU(2)$\cong S^3$. These vector fields may be defined by their action on functions $f : \text{SU}(2) \rightarrow \mathbb{C}$,

$$(D_J^L f)g := \frac{d}{dt}f(e^{-itJ}g)|_{t=0}, \quad (D_J^R f)g := \frac{d}{dt}f(ge^{itJ})|_{t=0}, \tag{15.2}$$

for all J in the LIE algebra su(2) of SU(2). Such vector fields form two commuting representations of su(2):

$$[D_J^\varepsilon, D_K^\varepsilon] = D_{[J,K]}^\varepsilon, \quad [D_J^L, D_K^R] = 0. \tag{15.3}$$

Here and in the following, the variable ε stands for L or R. The LAPLACE operator on S^3 is a second order differential operator that is built from the first order vector fields D_J^ε with the help of the KILLING form κ of su(2),

$$\Delta = \frac{1}{2}(\Delta^L + \Delta^R) \quad \text{with} \quad \Delta^\varepsilon = \kappa^{ab}D_{Ja}^\varepsilon D_{Jb}^\varepsilon, \tag{15.4}$$

where J^a is a basis of su(2) and κ_{ab} the associated KILLING matrix. Our goal in this first section is to diagonalize the Hamiltonian for a particle on S^3. Up to some simple constant factor, the eigenvalues of H coincide with those of the LAPLACIAN. Clearly, the eigenvalues of Δ must have some degeneracies. Note that the LAPLACE operator commutes with all the differential operators D_J^ε. Consequently, acting with the left and right invariant vector fields on some eigenfunction of Δ will result in another eigenfunction with the same eigenvalue. In other words, the eigenfunctions of Δ are organized into multiplets of the symmetry algebra su(2)$^L \oplus$ su(2)^R.

Before we continue our analysis of the LAPLACE operator, let us recall some facts about the representations of the LIE algebra su(2). While the results are probably familiar to most readers, our derivation is a little less standard. The LIE algebra su(2) is generated by three elements $\{J^0, J^\pm\}$ with the relations

$$[J^+, J^-] = 2J^0 \qquad [J^0, J^\pm] = \pm J^\pm . \tag{15.5}$$

Here, we shall be interested in finite dimensional representations. Any such representation contains a lowest weight vector, i.e. a special state $|j\rangle$ which is annihilated by J^-. Since J^- lowers the eigenvalue of J^0, the lowest weight vector is the eigenstate of J^0. We adopt conventions in which the smallest eigenvalue is given by $-j$:

$$J^0|j\rangle = -j|j\rangle , \qquad J^-|j\rangle = 0 .$$

By definition, the VERMA module V_j is the lowest weight representation that is generated from the vector $|j\rangle$ by application of J^+. In other words, the VERMA module V_j is spanned by states of the form

$$V_j = \text{span}\Big\{ |j; l\rangle := (J^+)^l|j\rangle \ \Big|\ l \geq 0 \Big\}. \tag{15.6}$$

The norm of these states can be computed in the same way as for states e.g. in free bosonic field theory. Using the fact that the annihilation operator J^- is the adjoint of the creation operator J^+ we can employ the commutation relations of the LIE algebra su(2) and the properties of the ground state $|j\rangle$ to show that

$$\langle j; l+1 | j; l+1 \rangle = (2j - l)(l+1) \langle j; l | j; l \rangle . \tag{15.7}$$

Inspection of the right-hand side shows that states with negative norm are bound to appear whenever $j \notin \mathbb{N}/2$. They are avoided if and only if $j \in \mathbb{N}/2$. For such half-integer values of j, the VERMA module V_j contains states of zero norm. These span an infinite dimensional subspace:

$$N_j = \text{span}\big\{ |j; l\rangle ; l \geq 2j+1 \big\} \subset V_j . \tag{15.8}$$

In order to construct a unitary representation of su(2) we remove the subspace N_j from the VERMA module V_j, i.e. we define

$$H_j := V_j/N_j, \qquad \dim H_j = 2j + 1. \tag{15.9}$$

By construction, the quotient space H_j carries a $(2j+1)$-dimensional representation of su(2). The representations we have obtained here are the usual finite dimensional irreducible representations of su(2) for spin j.

Armed with this knowledge about the representation of the left and right symmetries of our LAPLACIAN, we now return to the spectral decomposition of the LAPLACE operator on S^3. As we have emphasized before, the eigenspaces of the Laplacian form multiplets under the action of the left and right invariant vector fields. From the famous PETER-WEYL theorem we know how the space of functions $\mathcal{F}(\mathrm{SU}(2))$ on the 3-sphere decomposes with respect to the $\mathrm{su}(2)^L \oplus \mathrm{su}(2)^R$ symmetry:

$$\mathcal{F}(\mathrm{SU}(2))\big|_{\mathrm{su}(2)^L \oplus \mathrm{su}(2)^R} \cong \bigoplus_{j \in \frac{\mathbb{N}}{2}} H_j^L \otimes H_j^R . \tag{15.10}$$

The irreducible subspace $H_j^L \otimes H_j^R$ is spanned by the $(2j+1)^2$ elements of the representation matrices for the $(2j+1)$-dimensional representation of $\mathrm{SU}(2)$:

$$H_j^L \otimes H_j^R \cong \mathrm{span}\big\{ D_{mn}^j(g) \,\big|\, g \in \mathrm{SU}(2) ; \, m, n = -j, -j+1, \ldots, j \big\}. \tag{15.11}$$

We know that the LAPLACIAN acts on the elements of these representation matrices as

$$\Delta D_{mn}^j = j(j+1) D_{mn}^j . \tag{15.12}$$

Thereby, we have found the spectrum and the eigenfunctions of the LAPLACE operator on S^3: For each half-integer value of $j = 0, 1/2, 1, \ldots$ there exist $(2j+1)^2$ independent eigenfunctions with eigenvalue $j(j+1)$. This answers the question we had set out to study in this section.

15.2 The SU(2) WESS-ZUMINO-NOVIKOV-WITTEN Model

We now want to study a 2-dimensional conformal field theory in which the fields take values on the 3-sphere $S^3 \cong \mathrm{SU}(2)$. The action we are about to spell out involves an $\mathrm{SU}(2)$ valued field g on the 2-dimensional complex plane:

$$g : \begin{array}{c} \Sigma \cong \mathbb{C} \longrightarrow \mathrm{SU}(2) \\ (z, \bar{z}) \longmapsto g(z, \bar{z}) \in \mathrm{SU}(2) . \end{array} \tag{15.13}$$

To be concrete, we shall think of elements in the group $\mathrm{SU}(2)$ as 2×2 matrices of the form

$$g = \begin{pmatrix} \alpha & \beta \\ -\beta^* & \alpha^* \end{pmatrix} \quad \text{where} \quad |\alpha|^2 + |\beta|^2 = 1,$$

and both α and β are complex parameters. In the field theory, $\alpha = \alpha(z, \bar{z})$ and $\beta = \beta(z, \bar{z})$ become functions on the complex z-plane. Another parametrization of $\mathrm{SU}(2)$ group elements, one that we shall actually use a few times below, makes use of the exponential map. If $\sigma_a, a = 1, 2, 3$, denote the usual Pauli matrices, i.e., the representation matrices of the generators J_a in the 2-dimensional representation of $\mathrm{su}(2)$, then an element $g \in \mathrm{SU}(2)$ can be written in the form

$$g = e^{i\sigma_a X^a} \sim 1 + i\sigma_a X^a + \mathcal{O}(X^2) . \tag{15.14}$$

A sum over the index a is understood, and we omitted all terms in the expansion that contain higher powers of the parameters X^a. In the field theory, the real coefficients X^a get promoted to fields $X^a = X^a(z, \bar{z})$. Now we are prepared to present a first action for the field g:

$$S_0[g] := -\frac{k}{4\pi} \int_\Sigma d^2z \ \text{tr}\left(g^{-1}\partial g g^{-1}\bar{\partial} g\right)$$

$$\sim \frac{1}{2\pi\alpha'} \int d^2z \partial X^a \bar{\partial} X^b \kappa_{ab} + \mathcal{O}(X^3). \qquad (15.15)$$

On the right-hand side we have inserted the expansion (15.14) and displayed all terms that are quadratic in the fields X^a. The metric κ is given by $\kappa_{ab} = tr(\sigma_a\sigma_b)$, and we have replaced the parameter k by $k = 2/\alpha'$ so that the action has the same form as in eq. (2.20). Note, however, that the action (15.15) contains infinitely many terms of higher order in X^a that we did not display.

The action S_0 has many nice properties. In particular, it is invariant under global left/right translations by $h \in SU(2)$: $S_0[gh] = S_0[hg] = S_0[g]$. Nevertheless, S_0 has no direct use for string theory. In order to impose the physical state conditions of string theory, we need the quantum theory to possess VIRASORO symmetry. This is the case for all models for which the stress energy tensor of the quantum theory is traceless. In the introduction to Chapter 13 we had connected this feature of the world-sheet model to properties of the target space: To leading order in α', the stress energy tensor T is traceless if and only if the background satisfies the (super-)gravity equations of motion . The 3-sphere we are studying in this section is a background of constant non-vanishing curvature. In EINSTEIN's theory of gravitation, some non-vanishing energy density of matter is required as a source for curvature. Without the appropriate contributions from matter, a constant curvature background does not solve EINSTEIN's equations. Hence, the model (15.15) does not give rise to a quantum theory with VIRASORO symmetry.

Our discussion in the last paragraph also shows how to correct for the bug of S_0: we must add terms that arise from the presence of some matter fields with constant energy density. In models of closed strings, there is a variety of matter at our disposal. It includes the various form fields B, C_{p+1} and the dilaton. Here we shall try to satisfy the gravitational equations of motion with a non-trivial KALB-RAMOND field, partly because the latter is included in all string theories we have seen so far. The KALB-RAMOND field strength H is a 3-form whose 2-form potential we have denoted by B. The stress energy tensor of our model possesses vanishing trace if metric and B-field are related by

$$\beta_{ab} := \underbrace{R_{ab}}_{\text{Ricci tensor}} - \underbrace{\frac{1}{4}H_{acd}H_b{}^{cd}}_{H = dB} + \mathcal{O}(\alpha') = 0 .$$

In order to source the constant curvature metric on the 3-sphere with a RICCI tensor $R_{ab} = 2R^{-2}G_{ab}$, we need a constant field strength H. Any constant 3-form on the 3-sphere is proportional to the volume form:

$$\omega = tr\left(g^{-1}dg \wedge g^{-1}dg \wedge g^{-1}dg\right) \sim \epsilon_{abc}dX^a \wedge dX^b \wedge dX^c . \qquad (15.16)$$

Here, we inserted $[\sigma_a, \sigma_b] = i\epsilon_{abc}\sigma^c$ where ϵ is the 3-dimensional epsilon symbol. To formulate string theory in such a background with non-vanishing 3-form field strength $H \sim \omega$, we need a 2-form potential B such that $dB = H$. For the moment we will not worry about the existence of such a potential and simply denote it by $B = d^{-1}H$. Then, we introduce the new action:

$$S[g] = S_0[g] + \frac{ik}{12\pi} \int d^2z d^{-1}H \sim S_0[g] + \frac{ik}{12\pi} \int f_{abc}X^a dX^b \wedge dX^c$$

$$= S_0[g] + \frac{ik}{6\pi} \int d^2z f_{abc}X^a \partial X^b \bar{\partial} X^c . \qquad (15.17)$$

The new term is referred to as the WESS-ZUMINO (WZ) term. It describes the effect of a B field whose amplitude depends linearly on the coordinates X^a near the group unit. There are some problems with the previous formula, though. In fact, contrary to our assumption, the volume form ω on S^3 does not possess a globally defined 2-form potential. In order to provide a rigorous construction of S, we have to extend the field g from the 2-dimensional world-sheet Σ to a 3-dimensional domain Ω whose boundary is $\partial\Omega = \Sigma$. We shall denote the extension of g to Ω by $\tilde{g}$. Once such an extension has been chosen, we can employ STOKES's theorem to give a new definition of S:

$$S[g] = S_0[g] + \frac{ik}{12\pi} \int_\Omega tr\left(\tilde{g}^{-1}d\tilde{g} \wedge \tilde{g}^{-1}d\tilde{g} \wedge \tilde{g}^{-1}d\tilde{g}\right) . \qquad (15.18)$$

One may wonder whether the WESS-ZUMINO term in the action is a functional of the original field configuration g on Σ or rather of its extension $\tilde{g}$ to the 3-manifolds Ω. Clearly, there exist many such extensions $\tilde{g}$ of g from Σ to Ω. But most details of this extension do not contribute to the value of the WESS-ZUMINO term. Since the WESS-ZUMINO term involves the volume form on S^3, its value on $\tilde{g}$ measures only the volume of the image $\tilde{g}(\Omega)$ of Ω in SU(2)$\cong S^3$. The function g on Σ specifies the boundary of $\tilde{g}(\Omega)$, and hence the volume of $\tilde{g}(\Omega)$ is determined by g up to an integer multiple of the total volume $vol(S^3) = 12\pi$ of the 3-sphere. Consequently, while our action S is not uniquely specified by the function g, the phase $\exp(2\pi i S[g])$ is, provided we choose the parameter k to take integer values. From the FEYNMAN path integral prescription we can therefore conclude that $S[g]$ does specify a well-defined quantum field theory.

Having established $S[g]$ as a promising candidate for a conformal quantum field theory that describes strings moving on the 3-sphere, we can now study its properties. In particular, we can use the action $S[g]$ to derive the following classical equations of motion:

$$\underbrace{\bar{\partial}(g^{-1}\partial g) + \partial(g^{-1}\bar{\partial}g)}_{\text{from } S_0[g]} + \underbrace{\bar{\partial}(g^{-1}\partial g) - \partial(g^{-1}\bar{\partial}g)}_{\text{from WZ-term}} = 0. \tag{15.19}$$

The first two terms are easily derived from the original action $S_0[g]$. As the action itself, they are symmetric under exchange of the two world-sheet derivatives ∂ and $\bar{\partial}$. In the presence of the KALB-RAMOND field we obtain a second set of terms of a similar form, only that these are now anti-symmetric under the exchange of derivatives.[1] Adding the terms from the two types of contributions, we obtain

$$\bar{\partial}(g^{-1}\partial g) = 0. \tag{15.20}$$

This equation generalizes the corresponding wave equation $\partial\bar{\partial}X^a = 0$ for strings in flat space. Recall that the latter provided us with two sets of (anti-) holomorphic fields: the currents $J^a = i\partial X^a$ and $\bar{J}^a = i\bar{\partial}X^a$. The same is true for our theory on S^3. In fact, from eq. (15.19) we conclude that

$$-kg^{-1}\partial g =: J(z) = \sigma_a J^a(z) \tag{15.21}$$

is holomorphic. The following short computation shows that the theory also contains an anti-holomorphic counterpart $\bar{J}$:

$$0 = \bar{\partial}(g^{-1}\partial g) = -(g^{-1}\bar{\partial}gg^{-1})\partial g + g^{-1}\bar{\partial}\partial g = g^{-1}\partial(\bar{\partial}gg^{-1})g. \tag{15.22}$$

This suggests to define a second, anti-holomorphic set of currents by

$$k\bar{\partial}gg^{-1} =: \bar{J}(\bar{z}) = \sigma_a \bar{J}^a(\bar{z}). \tag{15.23}$$

As was explained in Chapter 13 the existence of such (anti-) holomorphic fields is the key to symmetry aided solutions of 2-dimensional conformal field theories. We shall now discuss how such a solution is constructed in the case at hand.

15.3 Solution of the WESS-ZUMINO-NOVIKOV-WITTEN Model

Since the components $J^a(z)$ and $\bar{J}^a(\bar{z})$ of the chiral currents $J(z)$ and $\bar{J}(\bar{z})$ are (anti-) holomorphic, we expand these fields in a Laurent series,

$$J^a(z) = \sum_{n\in\mathbb{Z}} J_n^a z^{-n-1}, \tag{15.24}$$

---

[1] One may easily see that the variation of the WESS-ZUMINO term contributes only through an integral over the boundary $\Sigma = \partial\Omega$.

and similarly for $\bar{J}^a(\bar{z})$. The modes J_n^a are the counterparts of the modes a_n^μ we have worked with for flat backgrounds. After quantization, these modes obey commutation relations that are somewhat reminiscent of the canonical commutation relations for the a_n^μ,

$$\left[J_n^a, J_m^b\right] = \mathrm{i}f^{ab}_{c}J_{m+n}^c + k\kappa_{ab}m\delta_{n+m,0}, \tag{15.25}$$

where f^{ab}_{c} are the structure constants of su(2) and κ_{ab} is the KILLING form, as before. While the second term is familiar from free bosonic field theory, the first term is new. It represents the effect of the curvature in the algebraic structure of the model. The above commutation relations define what is known as su(2) current algebra at level k. It is often denoted by $\widehat{\mathrm{su}}(2)_k$. From the generators of this algebra, we can also construct a copy of the VIRASORO algebra. Up to an overall normalization, we can simply copy the formulas we used in free field theory:

$$L_n := \frac{1}{2(k+2)} \sum_{m\in\mathbb{Z}} : J_{n-m}^a J_m^b : \kappa_{ab}. \tag{15.26}$$

This expression for the generators L_n is known as an affine SUGAWARA construction. The L_n may be shown to obey the usual VIRASORO commutation relations with the central charge

$$c_k = \frac{3k}{k+2}. \tag{15.27}$$

For the first time we see an example of a theory in which the central charge is not just proportional to the dimension of the background. In the SU(2) WESS-ZUMINO-NOVIKOV-WITTEN (WZNW) model the central charge satisfies $c_k \leq 3$, i.e., it is bounded from above by the dimension of S^3. The central charge approaches its largest value $c_{\max} = 3$ only in the limit $k \to \infty$ and can be as small as $c_1 = 1$ for $k = 1$. In this sense, the stringy S^3 may be considered as a space of fractal dimension. We shall see below that it really behaves a bit like a discrete space.

In order to use the affine current algebra symmetry to solve the model, we need some background from representation theory. In Chapter 13 we described the main ingredients of a symmetry-aided solution, namely the characters for irreducible highest weight representations along with their modular S and T matrices. We shall now describe these data for the su(2) current algebra. The lowest energy representations of the su(2) current algebra are those representations for which the L_0 eigenvalue is bounded from below. These representations are parametrized by the irreducible representations of the zero mode algebra su(2), i.e., they are labeled by a spin number $j \in \mathbb{N}/2$. But not all such values of j give rise to a unitary representation of the current algebra.

In order to determine further conditions on the allowed values of j, we follow the procedure we have tried out for the finite dimensional LIE algebra su(2) in the first section. Our first step is to construct VERMA modules $\mathcal{V}_j$. These are generated from ground states $|j; l\rangle \in H_j$ satisfying

$$J_n^a | j; l\rangle = 0 \text{ for } n > 0$$

by application of the creation operators J_n^a with $n < 0$. As we have seen before, such VERMA modules often contain negative norm states. Here we can find such negative norm states for all but a finite number of $j \in \mathbb{N}/2$.

Lemma 1: *The* VERMA *modules* $\mathcal{V}_j$ *contain vectors with negative norm if and only if* $j > k/2$.

Proof: For the proof of the Lemma it is crucial to observe that the three elements

$$H^+ = J_{-1}^+, \quad H^- = J_1^-, \quad H^0 = J_0^0 - \frac{k}{2}$$

satisfy the defining relations of the LIE algebra su(2). Thus, any VERMA module $\mathcal{V}_j$ carries an action of this su(2) algebra, in addition to the action of the zero modes. We have seen previously that representation of su(2) contains states with negative norm provided the eigenvalue of the operator H^0 on the lowest weight state is positive. On the ground states $|j; l\rangle \in H_j$ with $0 \leq l \leq 2j$, the element H^0 assumes the value $-j + l - k/2$. Consequently, there exist lowest weight states with positive eigenvalue of H^0 whenever $j > k/2$. Thus, the VERMA modules $\mathcal{V}_j$ with $j > k/2$ contain negative norm states. □

Hence, demanding the absence of negative norm states restricts us to the finite number of VERMA modules $\mathcal{V}_j$ with $j \leq k/2$. All these remaining VERMA modules contain zero norm states that must be removed in order to obtain unitary irreducible representations. The location of zero norm states can be found using the same kind of reasoning as in the proof of the previous lemma.

Lemma 2: *For $j \leq k/2$, the* VERMA *module* $\mathcal{V}_j$ *contains a multiplet of zero norm states in the representation H_{k+1-j} of the zero mode algebra. It is generated from the states*

$$\eta_k(j) := \left(J_{-1}^+ \right)^{k-2j+1} | j; 2j \rangle$$

by application of the zero modes J_0^a in the affine current algebra.

Proof: Our argument employs the same generators H^a that were introduced in the proof of the previous lemma. Let us pick the state $|j, 2j\rangle$, which is an eigenstate of J_0^0 with eigenvalue j. Hence, its eigenvalue under the action of H_0 is $j - k/2$. From the results in the first section we know that application of $\left(J_{-1}^+ \right)^{k-2j+1}$ will bring us to a state of zero norm. Under the action of J_0^0, the corresponding state $\eta_k(j)$ has eigenvalue $k - j + 1$. Using the lowering operator J_0^- we obtain the $2k - 2j + 3$ states in a representation H_{k-j+1} of the zero mode algebra su(2).

$\square$

Once we know the location of the vectors with zero norm, we can remove them to construct the irreducible representations

$$\mathcal{H}_j := \mathcal{V}_j / \mathcal{N}_j \quad \text{where} \quad \mathcal{N}_j \cong \mathcal{V}_{k-j+1}$$

is the submodule of $\mathcal{V}_j$ that is generated from the zero norm state $\eta_k(j)$ by application of the creation operators J_n^a with $n \leq 0$. The associated characters of the irreducible representations $\mathcal{H}_j$ are now easy to compute. The result reads

$$\chi_j(q) = \frac{q^{\frac{(j+1/2)^2}{k+2}}}{\eta^3(q)} \sum_{m \in \mathbb{Z}} (2j + 1 + 2m(k+2)) \, q^{m[2j+1+m(k+2)]} \tag{15.28}$$

$$= \frac{q^{\frac{(j+1/2)^2}{k+2}}}{\eta^3(q)} \left[(2j+1) - (2(k-j+1)+1)q^{k-2j+1} + \cdots \right]. \tag{15.29}$$

The first term $m = 0$ in this sum is the character of the full VERMA module over the $(2j + 1)$-dimensional space of ground states. On the ground states, the operator L_0 assumes the value $j(j+1)/(k+2)$. After subtracting the term $c/24 - 1/8$ we obtain the argument of the exponential that multiplies the entire character . The factor η^{-3} arises from applying three sets of creation operators, as usual, and it contains the factor $q^{-1/8}$. Since the construction of the irreducible representation space $\mathcal{H}_j$ involves removing zero norm states in the representation $\mathcal{V}_{k-j+1}$, we have to subtract the corresponding contributions to the characters . As we stated in Lemma 2, these states appear at level $k - 2j + 1$ above the ground state of the VERMA module $\mathcal{V}_j$. Removing the VERMA module $\mathcal{V}_{k-j+1}$ is achieved by the terms with $m = -1$. But this subtraction overdoes the job since $\mathcal{V}_{k-j+1}$ itself contains states that were not included in $\mathcal{V}_j$. These are organized in a VERMA module $\mathcal{V}_{j+k+2}$, and they are added back in by the terms with $m = 1$ etc.

The general theory of current algebras and their representations guarantees that the characters of the unitary positive energy representations close under modular transformations. The modular S matrix that we defined in eq. (13.11) may be shown to take the form

$$S_{ij} = \left(\frac{2}{k+2} \right)^{1/2} \sin \frac{(2i+1)(2j+1)}{k+2}. \tag{15.30}$$

The modular T transformation is determined by the conformal weight h_j of the ground states in $\mathcal{H}_j$. The matrix elements of the S matrix are real, and they satisfy

$\sum_j S_{ij} S_{jk} = \delta_{ik}$. This makes it easy for us to find at least one modular invariant among the possible combinations of characters from the left- and right-moving sector of the modes:

$$Z_k^{\text{WZNW}}(\tau) = \sum_{j=0}^{k/2} \delta_{ij} \, \chi_i(q) \, \chi_j(\bar{q}) \,. \tag{15.31}$$

As in the particle model, the spin of the left- and right-moving sector is the same for all the states that contribute. But there is an important difference with the particle theory. While the angular momentum j of a particle on S^3 is unbounded from above, it gets cut off by $k/2$ in the corresponding field theory. Such a cut-off means that wave functions of the field theory cannot resolve short distances in the target space. It appears as if the target space of the field theory is discrete or fuzzy.

Before we end this chapter, let us stress that the partition function (15.31) is not the only modular invariant one can build from the left and right characters of the su(2) current algebra. When the level takes one of the values $k = 4n + 2$, for example, another possibility is to combine left and right movers through

$$Z_{k;2}^{\text{WZNW}}(\tau) = \sum_{j=0}^{k/2} n^{ij} \chi_i(q) \, \chi_j(\bar{q}) \quad \text{where} \quad n^{ij} = \begin{cases} \delta_{i,j} & \text{for } j \in \mathbb{Z} \\ \delta_{\frac{k}{2}-i,j} & \text{for } j \in \mathbb{Z} + \frac{1}{2} \end{cases} \,. \tag{15.32}$$

This modular invariant can be obtained as a $\mathbb{Z}_2$ orbifold from the partition function (15.31) of the SU(2) WZNW model. It describes a model with target space $SU(2)/\mathbb{Z}_2 \cong SO(3)$.

Many more details on WZNW models, including extensive description of relevant results from the representation theory of affine Kac-Moody algebras and a comprehensive discussion of modular invariant partition functions, can be found e.g. in [18, 38].

Exercises

Problem 32. *Consider the representations of the finite dimensional* LIE *algebra* su(2) *we introduced and compute the characters*

$$\chi(z) := tr\left(z^{J^0}\right)$$

for the representations V_j, N_j, and H_j, which are defined in eqs. (15.6), (15.8), and (15.9), respectively.

Problem 33. *Show that $J^- |j, 2j + 1\rangle = 0$ where $|j, l\rangle := (J^+)^l |j\rangle$. Use the commutation relations (15.5) of the* su(2) *LIE algebra along with the properties $J^0 |j\rangle = -j|j\rangle$ and $J^- |j\rangle = 0$ of the "ground state" $|j\rangle$.*

Problem 34. *Consider the* su(2) *current algebra generated by* J_n^a *along with the* SUGAWARA *construction for the* VIRASORO *modes* L_n; *see eq. (15.26). The current algebra contains a* u(1) *subalgebra generated by the modes* J_n^0 *and the associated* VIRASORO *generators*

$$L_n' = \frac{1}{2} \sum_m : J_{n-m}^0 J_m^0 : .$$

Show that $\tilde{L}_n = L_n - L_n'$ *satisfy the relations of the* VIRASORO *algebra. What is the central charge?*

HINT: *You can use* $[L_n, J_m^0] = -m J_{n+m}^0 = [L_n', J_m^0]$, *which follows from the* SUGAWARA *constructions.*

16

CALABI-YAU Spaces

In the previous chapters we have learned some of the techniques from 2-dimensional quantum field theory that are needed to describe strings in curved backgrounds. For the next few Chapters we will turn our attention to the space-time aspects of strings moving in non-trivial compactifications. The next two chapters are devoted to CALABI-YAU compactifications, i.e., compactifications of 10-dimensional (type II) superstrings to theories with an extended 4-dimensional MINKOWSKI space-time.

As we have discussed in Chapter 10, type II superstring theories in 10 dimensions possess 32 supercharges. This is the maximal number of supercharges we could preserve when we compactify to four dimensions. It would correspond to N=8 supersymmetry of the compactified theory.[1] This maximal amount of supersymmetry is realized for torus compactifications. The CALABI-YAU compactifications we are about to discuss preserve eight supercharges, corresponding to an N=2 supersymmetry in four dimensions. In this case the background for our type IIA/B superstring theory has the form (13.1) with the external part being 4-dimensional MINKOWSKI space and some compact 6-dimensional CALABI-YAU surface $\mathscr{X}^6$. We will learn in some detail which properties of $\mathscr{X}$ are essential in order to guarantee 4-dimensional N=2 supersymmetry.

The following chapter is mainly devoted to some mathematical background. Without any attempt to motivate the structures from string theory, we shall introduce a certain class of complex manifolds with a Hermitian metric satisfying the so-called KÄHLER condition. Within this class we look for RICCI flat geometries, i.e., for metrics that satisfy the vacuum EINSTEIN's equations. Manifolds that are both KÄHLER and RICCI flat are known as CALABI-YAU manifolds. The last part of this chapter is devoted to some of their most important properties. We conclude by giving at least one non-trivial example. Many more details and references to the original literature can be found, e.g., in the lectures [11] or in the book [31].

---

[1] Recall that a MAJORANA spinor in four dimensions has four real components.

16.1 Complex Manifolds

A complex manifold $\mathscr{X}$ of complex dimension n is much like a real one, with one notable difference: that its charts (U_r, ϕ_r) take values in $\mathbb{C}^n$, i.e., $\phi_r : U_r \to \mathbb{C}^n$, and that the transition functions

$$\phi_r \circ \phi_s^{-1} : \phi_s(U_r \cap U_s) \to \mathbb{C}^n$$

are holomorphic. Thus, on a complex manifold one has a notion of (anti-)holomorphic forms, vector fields, etc.

Examples: The most trivial example of a complex manifold is of course $\mathbb{C}^n$. A more interesting case are the complex projective spaces $\mathbb{C}P^m$, which are defined as

$$\mathbb{C}P^m = \left\{ (w_1, \ldots, w_{m+1}) \in \mathbb{C}^m \setminus \{0\} \right\} / \sim \tag{16.1}$$

with the equivalence relation

$$(w_1, \ldots, w_{m+1}) \sim (w_1', \ldots, w_{m+1}') \Leftrightarrow (w_1, \ldots, w_{m+1}) = \lambda(w_1', \ldots, w_{m+1}')$$

for $\lambda \neq 0$. These spaces have real dimension $2m$, and a useful set of coordinate charts for them is

$$U_r := \left\{ w = (w_1, \ldots, w_{m+1}) \,\big|\, w_r \neq 0 \right\} / \sim, \tag{16.2}$$

$$\phi_r(w)^i = z_{(r)}^i = \begin{cases} \frac{w_i}{w_r} & \text{for } 1 \leq i < r \\ \frac{w_{i+1}}{w_r} & \text{for } r \leq i \leq m \end{cases}. \tag{16.3}$$

The transition functions $\phi_s \circ \phi_r^{-1}$ are simply given by multiplication with w_r/w_s so that they are holomorphic. The lowest dimensional complex projective space $\mathbb{C}P^1$ is the 2-dimensional sphere S^2.

On a complex manifold we can introduce the space $\Omega^{p,q}(\mathscr{X})$ of (p, q) forms. Locally, i.e., in any given chart, such forms may be written as

$$w = w_{i_1 \cdots i_p, \bar{\imath}_1 \cdots \bar{\imath}_q} \, dz^{i_1} \wedge \cdots \wedge dz^{i_p} \wedge d\bar{z}^{\bar{\imath}_1} \wedge \cdots \wedge d\bar{z}^{\bar{\imath}_q} . \tag{16.4}$$

The total differential d can be decomposed into a holomorphic and an anti-holomorphic part $d = \partial + \bar{\partial}$ with

$$\partial : \Omega^{p,q}(\mathscr{X}) \to \Omega^{p+1,q}(\mathscr{X}), \quad \bar{\partial} : \Omega^{p,q}(\mathscr{X}) \to \Omega^{p,q+1}(\mathscr{X}) . \tag{16.5}$$

In local charts, we can use the following explicit expressions:

$$\partial = \sum_{i=1}^{n} dz^i \frac{\partial}{\partial z^i}, \quad \bar{\partial} = \sum_{\bar{i}=1}^{n} d\bar{z}^{\bar{i}} \frac{\partial}{\partial \bar{z}^{\bar{i}}}. \tag{16.6}$$

Having introduced the spaces $\Omega^{p,q}(\mathscr{X})$ and the operators ∂ and $\bar{\partial}$, it is natural to investigate the corresponding cohomology classes, i.e. the equivalence classes of closed modulo exact forms:

$$H_{\partial}^{p,q}(\mathscr{X}) = \frac{\ker\left(\partial : \Omega^{p,q}(\mathscr{X}) \to \Omega^{p+1,q}(\mathscr{X})\right)}{\mathrm{Im}\left(\partial : \Omega^{p-1,q}(\mathscr{X}) \to \Omega^{p,q}(\mathscr{X})\right)},$$

$$H_{\bar{\partial}}^{p,q}(\mathscr{X}) = \frac{\ker\left(\bar{\partial} : \Omega^{p,q}(\mathscr{X}) \to \Omega^{p,q+1}(\mathscr{X})\right)}{\mathrm{Im}\left(\bar{\partial} : \Omega^{p,q-1}(\mathscr{X}) \to \Omega^{p,q}(\mathscr{X})\right)}.$$

The HODGE numbers $h^{p,q}$ are defined to be the dimensions of $H_{\partial}^{p,q}$:

$$h^{p,q} = dim(H_{\partial}^{p,q}), \quad \bar{h}^{p,q} = dim(H_{\bar{\partial}}^{p,q}). \tag{16.7}$$

The numbers $h^{p,q}$ and $\bar{h}^{p,q}$ are related through complex conjugation: $h^{p,q} = \bar{h}^{q,p}$. For a complex 3-dimensional manifold, the HODGE numbers are often arranged in the following diamond shaped array:

$$
\begin{array}{ccccccc}
 & & & h^{3,3} & & & \\
 & & h^{3,2} & & h^{2,3} & & \\
 & h^{3,1} & & h^{2,2} & & h^{1,3} & \\
h^{3,0} & & h^{2,1} & & h^{1,2} & & h^{0,3} \\
 & h^{2,0} & & h^{1,1} & & h^{0,2} & \\
 & & h^{1,0} & & h^{0,1} & & \\
 & & & h^{0,0} & & &
\end{array}
\tag{16.8}
$$

It is usually referred to as the HODGE diamond of the complex manifold $\mathscr{X}^6$. There is another diamond that contains the numbers $\bar{h}^{p,q}$ that is related to the one we have displayed by a reflection along the vertical axis.

16.2 Hermitian Manifolds

Any string background $\mathscr{X}$ must come equipped with a metric $g : T\mathscr{X} \otimes T\mathscr{X} \to \mathbb{R}$. In a given coordinate patch of a complex manifold the most general metric takes the form

$$g = g_{ij} dz^i \otimes dz^j + g_{\bar{i}\bar{j}} d\bar{z}^{\bar{i}} \otimes d\bar{z}^{\bar{j}} + g_{i\bar{j}} d\bar{z}^{\bar{i}} \otimes dz^j + g_{i\bar{j}} dz^i \otimes d\bar{z}^{\bar{j}}. \tag{16.9}$$

We will not consider general complex RIEMANNIAN geometries, but restrict ourselves to those for which $g_{ij} = 0 = g_{\bar{i}\bar{j}}$:

$$g = g_{\bar{i}j} \, d\bar{z}^{\bar{i}} \otimes dz^j + g_{i\bar{j}} \, dz^i \otimes d\bar{z}^{\bar{j}} \, . \tag{16.10}$$

A metric of this form is known as *Hermitian*. It provides us with a pairing between holomorphic and anti-holomorphic indices. When we lower or raise holomorphic indices they become anti-holomorphic and vice versa.

On spaces with a Hermitian metric, there exists an important re-interpretation of the HODGE numbers. This needs a bit of preparation. Given some metric g in $\mathscr{X}$, we can construct a canonical volume form Ω_{vol}. We use this volume form to introduce a linear map

$$\star : \quad \Omega^{p,q}(\mathscr{X}) \quad \rightarrow \quad \Omega^{n-p,n-q}(\mathscr{X}) \tag{16.11}$$

such that

$$v \wedge \star w = (v, w) \, \Omega_{\text{vol}} \tag{16.12}$$

for all elements $w, v \in \Omega^{p,q}(\mathscr{X})$ and $n = dim_{\mathbb{C}} \mathscr{X}$. As usual, (v, w) denotes the inner product of (p, q) forms. In local coordinates it is given by

$$(v, w) = \bar{v}_{\bar{i}_1 \ldots \bar{i}_p, i_1 \ldots i_q} w_{j_1 \ldots j_p, \bar{j}_1 \ldots \bar{j}_q} \, g^{\bar{i}_1 j_1} \ldots g^{i_q \bar{j}_q} \, .$$

The map $\star$ is known as the HODGE $\star$ operation. Next let us consider the following combination of the holomorphic differential ∂ and the HODGE $\star$ operation:

$$\partial^\dagger := - \star \partial \star : \Omega^{p,q}(\mathscr{X}) \rightarrow \Omega^{p-1,q}(\mathscr{X}) \, . \tag{16.13}$$

Note that $\partial^\dagger$ lowers the holomorphic degree p of a (p, q) form. Similarly, one also introduces a map $\bar{\partial}^\dagger$ that reduces the anti-holomorphic degree q of a differential form on $\mathscr{X}$ by one unit. From ∂ and $\partial^\dagger$, we can build an interesting map Δ that operates within the space of (p, q) forms:

$$\Delta := \partial \partial^\dagger + \partial^\dagger \partial : \Omega^{p,q}(\mathscr{X}) \mapsto \Omega^{p,q}(\mathscr{X}) \, . \tag{16.14}$$

Now we have all the ingredients to state the so-called complex HODGE theorem. It identifies the HODGE numbers $h^{p,q}$ as the dimension of the kernel of Δ:

$$h^{p,q} = dim\{w \in \Omega^{p,q} \, | \, \Delta w = 0\} \, .$$

In other words, $h^{p,q}$ gives the number of (p, q) *harmonic forms*, i.e., of zero modes of the LAPLACIAN Δ in the space of (p, q) forms. This new characterization of the HODGE numbers will become very useful in the next chapter. Moreover, since the $\star$ map is an isomorphism $\star : \Omega^{p,q} \rightarrow \Omega^{n-p,n-q}$ and Δ commutes with $\star$, we conclude that $h^{p,q} = h^{n-p,n-q}$. In other words, the HODGE diamonds of a Hermitian complex manifold are invariant under reflection with respect to the horizontal axis.

16.3 KÄHLER Manifolds

After these more general comments on complex spaces with Hermitian metric g we are going to narrow things by considering only those metrics that satisfy the so-called KÄHLER condition. To state this property of g, we note that the coefficients of a Hermitian metric can be used to build the following form:

$$\omega := ig_{i\bar{j}}\, dz^i \wedge d\bar{z}^{\bar{j}}\,. \tag{16.15}$$

This $(1, 1)$ form ω is the central object in the definition of a KÄHLER manifold. A complex manifold $\mathscr{X}$ with the Hermitian metric g and associated form ω is called a KÄHLER *manifold* if ω is closed, i.e. if $d\omega = 0$, where $d = \partial + \bar{\partial}$. If this condition is satisfied, the form ω is referred to as a KÄHLER form.

Before we proceed, let us spell out the KÄHLER condition in terms of the components $g_{i\bar{j}}$. By acting with the total differential d in ω and comparing coefficients we conclude that

$$\partial_k g_{i\bar{j}} = \partial_i g_{k\bar{j}}, \quad \partial_{\bar{k}} g_{i\bar{j}} = \partial_{\bar{j}} g_{i\bar{k}}\,. \tag{16.16}$$

From these equations we may infer that a KÄHLER metric possesses a local potential, i.e. in every chart, $g_{i\bar{j}}$ may be obtained by taking derivatives of some function $K(z, \bar{z})$:

$$g_{i\bar{j}}^{(r)} = \frac{\partial^2 K^{(r)}(z, \bar{z})}{\partial z^i \partial \bar{z}^{\bar{j}}}\,. \tag{16.17}$$

$K(z, \bar{z})$ is known as the local KÄHLER potential. Obviously, the converse is also true, i.e. if the metric g can be obtained from a potential K, then is satisfies the KÄHLER condition. For the time being, we will not attempt to motivate the KÄHLER condition from string theory. We simply treat it as an interesting condition that mathematicians like to work with. Our aim in this chapter is to study some of its consequences. We will return to the origin of the condition in the next chapter.

Examples: The simplest example of a KÄHLER manifold is $\mathbb{C}$ together with the metric $g = dz \otimes d\bar{z}$ and potential $K(z, \bar{z}) = |z|^2$. This one is a bit too simple to be of real interest to us.

A more relevant example is provided by the complex projective spaces $\mathbb{CP}^m$ with the FUBINI-STUDY metric g. Their KÄHLER metric comes from a potential K that we can easily write on each of the local coordinate patches U_r:

$$K^{(r)} = \log\left(\sum_{i=1}^{m} \left|z_{(r)}^i\right|^2 + 1\right) = \log \sum_{i=1}^{m+1} \left|\frac{w_i}{w_r}\right|^2\,, \tag{16.18}$$

$$g^{(r)} = \frac{\delta_{i\bar{\imath}}\left(\sum_{j=1}^{m}\left|z_{(r)}^{j}\right|^{2}+1\right) - z_{(r),\bar{\imath}}\bar{z}_{(r),i}}{\left(\sum_{j=1}^{m}\left|z_{(r)}^{j}\right|^{2}+1\right)^{2}}\, dz_{(r)}^{i}\otimes d\bar{z}_{(r)}^{\bar{\imath}}\,. \tag{16.19}$$

KÄHLER manifolds possess a number of important features that we shall exploit extensively in the next chapter. The first feature we would like to mention concerns the harmonic theory. In the previous section we introduced the two LAPLACIANS Δ and $\bar{\Delta}$. The latter was defined as in (16.14) but with $\bar{\partial}$ replacing ∂. On a general Hermitian manifold, there is no reason for Δ and $\bar{\Delta}$ to agree. But on a KÄHLER manifold one finds that

$$\Delta = \partial\partial^{\dagger} + \partial^{\dagger}\partial = \bar{\partial}\bar{\partial}^{\dagger} + \bar{\partial}^{\dagger}\bar{\partial} = \bar{\Delta}\,.$$

From the complex HODGE theorem we conclude that $h^{p,q} = \bar{h}^{p,q}$. With the help of the fundamental property $\bar{h}^{p,q} = h^{q,p}$ we infer $h^{p,q} = h^{q,p}$. Hence, the HODGE diamond of a KÄHLER manifold is invariant under reflection with respect to the vertical axis.

We conclude our short discussion of KÄHLER manifolds with a few formulas characterizing their differential geometry. One of the most basic quantities of a RIEMANNIAN manifold is the LEVI-CIVITA connection:

$$\Gamma^{\kappa}{}_{\mu\rho} := \frac{1}{2}g^{\kappa\lambda}\left(\frac{\partial g_{\lambda\rho}}{\partial x^{\mu}} + \frac{\partial g_{\mu\lambda}}{\partial x^{\rho}} - \frac{\partial g_{\mu\rho}}{\partial x^{\lambda}}\right)\,. \tag{16.20}$$

The CHRISTOFFEL symbols $\Gamma^{\kappa}{}_{\mu\rho}$ may be considered as matrix element of the metric connection Γ_{μ} that describes how tangent vectors are rotated under parallel transport along the coordinate x^{μ}. On a KÄHLER manifold, the formula for the LEVI-CIVITA connection simplifies considerably:

$$\Gamma^{l}_{jk} = g^{l\bar{s}}\frac{\partial g_{k\bar{s}}}{\partial z^{j}}\,, \qquad \Gamma^{\bar{l}}_{\bar{j}\bar{k}} = g^{\bar{l}s}\frac{\partial g_{\bar{k}s}}{\partial z^{\bar{j}}}\,. \tag{16.21}$$

All other CHRISTOFFEL symbols disappear. In particular, parallel transport on a KÄHLER manifold preserves the complex structure.

It will be very important for us to determine the curvature and the RICCI tensor of a KÄHLER manifold. For a general RIEMANNIAN manifold, the curvature tensor is obtained through the following prescription:

$$R^{\kappa}{}_{\lambda\mu\nu} = \partial_{\mu}\Gamma^{\kappa}{}_{\nu\lambda} - \partial_{\nu}\Gamma^{\kappa}{}_{\mu\lambda} + \Gamma^{\kappa}{}_{\mu\eta}\Gamma^{\eta}{}_{\nu\lambda} - \Gamma^{\kappa}{}_{\nu\eta}\Gamma^{\eta}{}_{\mu\lambda}\,. \tag{16.22}$$

We also recall that the RICCI tensor $R_{\mu\nu}$ is obtained from the curvature tensor by tracing over the indices κ and λ: $R_{\mu\nu} = R^{\kappa}{}_{\kappa\mu\nu}$. We can now insert our formula for the CHRISTOFFEL symbols on a KÄHLER manifold into the general expression for the associated curvature and RICCI tensors to obtain

$$R^{\bar{k}}{}_{\bar{l}ij} = -\frac{\partial \Gamma^{\bar{k}}_{\ \bar{i}\bar{l}}}{\partial z^j}, \quad R_{\bar{i}j} = -\frac{\partial \Gamma^{\bar{k}}_{\ \bar{i}k}}{\partial z^j}. \tag{16.23}$$

The reader is invited to compute the CHRISTOFFEL symbols and the RICCI tensor for the FUBINI-STUDY metric on complex projective spaces.

16.4 CALABI-YAU Manifolds

Let us continue to accept that the string backgrounds we want to use are to be found within the class of KÄHLER manifolds. As we have mentioned several times throughout the last few chapters, superstrings can move only in geometries that satisfy the equations of motion of 10-dimensional supergravity. If we furthermore require the absence of matter, then this means that the metric needs to be RICCI flat. In the mathematics literature, RICCI flat KÄHLER manifolds have been dubbed CALABI-YAU spaces.

This name originates from one of their important properties that was first conjectured by Calabi and later proven by Yau. According to their fundamental result, a compact complex space $\mathscr{X}$ possesses a KÄHLER metric g with vanishing RICCI tensor if and only if the first CHERN class $c_1(\mathscr{X})$ vanishes. Locally the first CHERN class of $\mathscr{X}$ is given by

$$c_1(\mathscr{X}) = R_{i\bar{j}} \, dz^i \wedge d\bar{z}^j \in H^{1,1}(\mathscr{X}).$$

This element of the cohomology group $H^{1,1}$ clearly vanishes if the RICCI tensor does. What is difficult to prove is the other direction: that a compact KÄHLER manifold with a vanishing first CHERN class admits a KÄHLER metric in the same class with vanishing RICCI curvature.

While the characterization of CALABI-YAU manifolds through the vanishing of the first CHERN class is very useful in finding examples of such manifolds, there is another characterization that we shall exploit extensively below. It is possible to show that sufficiently generic CALABI-YAU manifolds can be recognized by their HODGE diamond. In fact, under certain technical conditions that won't be relevant for us, the HODGE diamond of a CALABI-YAU manifold has the following special form:

$$
\begin{array}{ccccccc}
 & & & 1 & & & \\
 & & 0 & & 0 & & \\
 & 0 & & h^{1,1} & & 0 & \\
1 & & h^{2,1} & & h^{2,1} & & 1 \\
 & 0 & & h^{1,1} & & 0 & \\
 & & 0 & & 0 & & \\
 & & & 1 & & &
\end{array}
\tag{16.24}
$$

Examples: The simplest KÄHLER manifolds that we considered before, namely $\mathbb{C}^n$ and $\mathbb{CP}^m$, are not CALABI-YAU, the first because it is not compact and the second because its RICCI tensor is not zero. However, it turns out that some submanifolds of $\mathbb{CP}^m$ are. These are given by the sets of zeros of some carefully selected homogeneous polynomials. More precisely if

$$\mathscr{X} = \left\{ z = [w_1, \ldots, w_{m+1}] \in \mathbb{CP}^m \,\middle|\, P_i(z) = 0 \text{ for } i = 1, \ldots, N \right\} \quad (16.25)$$

with

$$P_s(\lambda w_1, \ldots, \lambda w_{m+1}) = \lambda^{l_s} P_s(w_1, \ldots, w_{m+1}) \quad \text{and} \quad m + 1 = \sum_{s=1}^{N} l_s, \quad (16.26)$$

then $\mathscr{X}$ is a CALABI-YAU manifold. Its metric is obtained by restricting the FUBINI-STUDY metric to $T\mathscr{X} \otimes T\mathscr{X}$. Manifolds of this type are known as complete intersection CALABI-YAU spaces.

The very simplest example of a CALABI-YAU threefold from this family is obtained from $m = 4$ and $N=1$, so that $l_s = 5$. To be specific, let us choose the following homogeneous polynomial of degree five:

$$P = \sum_{i=1}^{5} w_i^5 . \quad (16.27)$$

Its zeroes define a 3-dimensional subset of $\mathbb{CP}^4$. The resulting CALABI-YAU manifold is called the quintic. If we take $m = 5$, then in order to obtain a manifold with three complex dimensions, we need to have $N=2$, so that we have $l_1 + l_2 = 6$. In the case where one of the l_i is $l_i = 1$ we can eliminate the associated coordinate and work with a smaller value of m. Therefore, the only genuinely new CALABI-YAU threefolds we obtain at $m = 5$ are $l_1 = l_2 = 3$ or $l_1 = 2, l_2 = 4$ etc.

CALABI-YAU manifolds actually come in continuous families that share the same HODGE diamond. In the class of complete intersection CALABI-YAU spaces, the relevant parameters are related to the coefficients of the polynomials P_s. As an example we shall treat the quintic $m = 4, l_1 = 5$. The number of continuous parameters that describe the possible CALABI-YAU manifolds is given by the number of different homogeneous monomials in five variables of degree five. Generally, if the degree is l and the number of variables is $m + 1$, then the number of different monomials is counted by $(l+m)! / l! \, m!$, which – for the quintic – gives 126 possible parameters. Since there are $25 - 1$ possible linear redefinitions of the coordinates, corresponding to the group PGL(5), the actual number of independent parameters is $126 - 24 = 102$.

The number of continuous parameters of a CALABI-YAU manifold is actually given by the sum of its non-trivial HODGE numbers $h^{2,1}$ and $h^{1,1}$. Since the KÄHLER form is an element of $H_{\bar\partial}^{1,1}$, the number $h^{1,1}$ counts possible deformations of the KÄHLER structure. Additionally, a form in $H_{\bar\partial}^{2,1}$ can give rise to a deformation of the metric, with

$$g_{ij} = w_{i\bar k\bar l}\,\Omega_j^{\bar k\bar l}, \quad g_{\bar i\bar j} = w_{\bar i kl}\,\Omega_{\bar j}^{kl}, \tag{16.28}$$

where Ω_{ijk}, respectively $\Omega_{\bar i\bar j\bar k}$, are the non-trivial elements in $H_{\bar\partial}^{3,0}$ and $H_{\bar\partial}^{0,3}$, respectively. After such a deformation, the metric is no longer of the form (16.10). But it may be brought back into this form by changing the complex structure on the manifold. In this way, the elements of $H^{2,1}$ and $H^{1,2}$ generate variations of the complex structure. For the quintic one finds $h^{2,1} = 101$ and $h^{1,1} = 1$. The sum of these two numbers agrees with the total number of parameters of the quintic CALABI-YAU.

Exercises

Problem 35. *The complex projective space* $\mathbb{CP}^1$ *with the usual* FUBINI-STUDY *metric is an example of a complex* KÄHLER *manifold. On the northern hemisphere, its metric can be obtained from the* KÄHLER *potential*

$$K(z,\bar z) = \log(z\bar z + 1)\,.$$

Determine the CHRISTOFFEL *symbols and curvature and show that* $\mathbb{CP}^1$ *is not a* CALABI-YAU *manifold.*

Problem 36. *Determine the dimension of the moduli space of the unique 2-dimensional complete intersection* CALABI-YAU *space that can be obtained by imposing a single equation in* $\mathbb{CP}^3$.

17

CALABI-YAU Compactifications

In the last chapter we have sketched some results from mathematics. We looked at complex manifolds with Hermitian metric and then imposed the KÄHLER condition (16.16) without any further motivation. After combining the KÄHLER condition with RICCI flatness we ended up with the class of CALABI-YAU manifolds. The results of mathematics ensure that the HODGE diamond of a CALABI-YAU manifold has a very special form. It contains only two non-trivial entries, $h^{1,1} = h^{2,2}$ and $h^{1,2} = h^{2,1}$. All other entries are either zero or one; see Figure 16.24. Let us finally also recall that the numbers $h^{p,q}$ count the number of harmonic (p, q) forms on the manifold.

We made very little attempt to connect the KÄHLER condition and its various consequences to our goal, namely to construct 4-dimensional string compactifications with N=2 supersymmetry in space-time. The aim of the present chapter is to fill this gap. In the first half we shall show that the massless spectrum of type II compactifications on CALABI-YAU manifolds is organized in multiplets of the 4-dimensional N=2 POINCARÉ superalgebra. The second part is devoted to a deeper understanding of the KÄHLER condition. Our strategy is to connect this condition to the GSO projection in curved backgrounds.

17.1 CALABI-YAU Spaces and 4D Supersymmetry

We now want to investigate how the 4-dimensional low-energy effective field theories depend on the CALABI-YAU space we compactify our type IIA/B string theory on. As we mentioned before, the resulting gravity theories come with an N=2 supersymmetry in four dimensions, i.e. they possess eight supercharges. Our main goal is to determine the type and number of massless N=2 supermultiplets that appear in a given compactification.

This needs a bit of preparation. To begin, let us recall the bosonic content of the relevant supermultiplets. As in all massless supermultiplets, half of the supercharges decouple. Hence, we are left with four supercharges that we combine into two creation and two annihilation operators. The two fermionic creation operators carry helicity $\lambda = 1/2$. With this in mind we can easily determine the bosonic content of N=2 supermultiplets. We shall need

1. the *gravity multiplet*, which contains a tensor field T and a vector field V. This multiplet is generated from the vector with helicity $\lambda_V = 1$ by application of the two creation operators. The only other bosonic state is obtained when we apply two creation operators. It is a state of helicity $\lambda = 2$. Along with the conjugate states of helicity $\lambda = -1$ and $\lambda = -2$, the space of bosonic fields contains the two polarizations $\lambda_T = \pm 2$ of a massless graviton multiplet T. They come combined with the two polarizations $\lambda = \pm 1$ of a massless vector multiplet in four dimensions.
2. the *vector multiplet*, which contains two scalar fields S and one vector field V. It is generated from a scalar with helicity $\lambda = 0$ and hence contains another bosonic state of helicity $\lambda = 1$. Once we add the conjugate multiplet, we obtain the two polarizations $\lambda_V = \pm 1$ of a massless vector V and two scalars.
3. the *hyper multiplet* containing four scalars S. It can be generated from the two states of helicity $\lambda = -1/2$ in a massless Dirac multiplet. After application of the two creation operators with helicity $\lambda = 1/2$ we obtain four scalar fields with helicity $\lambda = 0$.

There is another key idea we need to explain before discussing the massless spectrum of CALABI-YAU compactifications: the KALUZA-KLEIN reduction. Let us consider the very simplest version of this reduction in which we start from a field theory in a $(D + d)$-dimensional space-time of the form $\mathbb{R}^{1,D-1} \times \mathscr{X}^d$. On this space-time we place some bosonic multiplet $\Phi_A = \Phi_A(x, \theta)$ satisfying the massless KLEIN-GORDON equation:

$$\left(-\frac{\partial^2}{\partial x_0^2} + \sum_{i=1}^{D-1} \frac{\partial^2}{\partial x_i^2} + \Delta_\theta \right) \Phi_A(x, \theta) = 0. \tag{17.1}$$

The subscript A labels different components of the multiplet, θ denotes coordinates on the compact surface $\mathscr{X}^d$, and Δ_θ is the LAPLACIAN. In the case where $\mathscr{X}^d$ is a d-dimensional torus with angular coordinates $\theta_a, a = 0, \ldots, d$, we have

$$\Delta_\theta = \sum_{a=1}^{d} \frac{\partial^2}{\partial \theta_a^2}.$$

Whatever the LAPLACIAN looks like, it possesses a basis of eigenfunctions $\phi_n(\theta)$ with eigenvalues $-\lambda_n$. This basis can be used to expand the fields $\Phi_A(x, \theta)$:

$$\Phi_A(x, \theta) = \sum_n \phi_n(\theta) \, \Phi_A^{(n)}(x). \tag{17.2}$$

When plugged back into the KLEIN-GORDON equation we obtain

$$\left(-\frac{\partial^2}{\partial x_0^2} + \sum_{i=1}^{D-1} \frac{\partial^2}{\partial x_i^2} - \lambda_n\right) \Phi_A^{(n)}(x) = 0. \tag{17.3}$$

What we obtain is an infinite set of multiplets $\Phi_A^{(n)}$ in the D-dimensional MINKOWSKI space $\mathbb{R}^{1,D-1}$ with masses $M_n^2 = \lambda_n$. The only massless multiplets arise from those eigenfunctions ϕ_n of the LAPLACIAN Δ_n whose eigenvalue λ_n vanishes. Hence, in order to find the massless multiplets we have to study the zero modes of the LAPLACIAN on the compactification manifold $\mathscr{X}^d$.

Now we can investigate the massless spectrum of type IIA supergravity compactified on a CALABI-YAU space with HODGE numbers $h^{1,1}$ and $h^{1,2}$. The results of the analysis are summarized in Table 17.1. We will explain a few of the entries. We recall that type IIA supergravity contains the metric G, the KALB-RAMOND potential B, the dilaton ϕ, and two $p + 1$-form potentials C_1 and C_3. In order to illustrate how we find the massless multiplets of the compactified model we shall look at the 3-form potential C_3. Physical polarizations are described by a 3-form in eight transverse directions that are spanned by two coordinates $x^\alpha, \alpha = 2, 3$, of the MINKOWSKI (or external) space and six directions $z^i, \bar{z}^j$ of the compact (or internal) manifold $\mathscr{X}^6$. We can expand such a form in a basis of 3-forms,

$$C_3 \sim \sum_k C_{3;k}^{(1)}(x)\, dx^2 \wedge dx^3 \wedge \omega_k^{(1)} + \sum_m C_{3;\alpha,m}^{(2)}(x)\, dx^\alpha \wedge \omega_m^{(2)} + \sum_n C_{3;n}^{(3)}(x)\, \omega_n^{(3)}, \tag{17.4}$$

where $\omega^{(r)}$ are forms of total degree $r = p + q$ on the compactification manifold $\mathscr{X}^6$ and the sums run over a complete basis of such forms. As we explained above, massless fields are associated with zero modes of the LAPLACIAN Δ on the compactification manifold. The LAPLACIAN in the space of all r-forms was constructed in eq. (16.14). In the case of a CALABI-YAU manifold, this LAPLACIAN has no zero modes in the space of 1-forms. Hence, there are no contributions to the massless spectrum from the first term in eq. (17.4). For 2- and 3-forms, the outcome is different. In fact, there are $h^{1,1}$ zero modes of Δ in the space of 2-forms. These give rise to $h^{1,1}$ vector fields $C_{3;\alpha,m}^{(1,1)}(x)$. In the space of 3-forms there are two zero modes from the space of $(3,0)$- and $(0,3)$-forms. In addition, we obtain $2h^{1,2} = 2h^{2,1}$ zero modes from $(2,1)$- and $(1,2)$-forms. All these correspond to scalars in the 4-dimensional world. The results of our short analysis are summarized in the last line of Table 17.1. The other lines can be obtained by a similar analysis. Most of them are much simpler than the analysis for C_3, so we leave them as

Table 17.1 *Field content of type IIA Calabi-Yau compactification*

IIA	$(0,0)$	$(1,1)$	$(1,2)$ $(2,1)$	$(0,3)$ $(3,0)$
G	T	$h^{1,1}S$	$2h^{1,2}S^*$	$-$
B	S	$h^{1,1}S$	$-$	$-$
ϕ	S	$-$	$-$	$-$
C_1	V	$-$	$-$	$-$
C_3	$-$	$h^{1,1}V$	$2h^{1,2}S$	$2S$

an exercise. The only entry that requires some additional comments appears in the first line for the graviton G. The metric G has two indices. When these are both in the external directions, we obtain the polarizations of a graviton in the 4-dimensional MINKOWSKI space. Alternatively, the indices can also be internal. Having one holomorphic and one anti-holomorphic index provides $h^{1,1}$ scalars. But in addition the two indices of G can be both holomorphic or both anti-holomorphic. As we argued at the end of the previous chapter, such components of the metric may be mapped to $(2,1)$- or $(1,2)$-forms with the help of the unique harmonic $(3,0)$- or $(0,3)$-form, respectively. This explains why we put $2h^{1,2}S$ in the top line of the third column of Table 17.1.

We can now combine the scalars that appear in the third column to form $h^{1,2}$ hyper-multiplets, and then take all the fields in the second column and combine them to $h^{1,1}$ vector multiplets. Afterwards, we see that the tensor and the vector field appearing in the first column form a gravity multiplet, which leaves only two scalars in the first column and two in the last, which give exactly one last hyper-multiplet. Our final conclusion for the massless content of type IIA theory on a CALABI-YAU three-fold is summarized as

$$h^{1,1} \quad \text{vector multiplets}$$

$$\text{IIA:} \quad h^{1,2} + 1 \text{ hyper-multiplets}$$

$$1 \quad \text{gravitational multiplet.}$$

The same analysis can then be performed for type IIB supergravity. We have written its field content in the left-most column of the table below. Most rows can be read directly from our Table 17.1 for massless multiplets in IIA theory. Note that the form fields C_0 and C_2 behave like ϕ and B, respectively.

The only new row is the last one for the self-dual 4-form field C_4^{sd}. Since self-duality removes half of the physical polarizations, we can perform the analysis for

Table 17.2 *Field content of type IIB Calabi-Yau compactification*

IIB	(0,0)	(1,1)	(1,2) (2,1)	(0,3) (3,0)
G	T	$h^{1,1}S$	$2h^{1,2}S$	$-$
B	S	$h^{1,1}S$	$-$	$-$
ϕ	S	$-$	$-$	$-$
C_0	S	$-$	$-$	$-$
C_2	S	$h^{1,1}S$	$-$	$-$
C_4^{sd}	$-$	$h^{1,1}S$	$h^{1,2}V^*$	V^*

a 4-form without self-duality condition and then divide the outcome by two. There is no entry in the first column because we cannot make a 4-form out of dx^2 and dx^3. The second entry arises from $h^{1,1}$ 2-forms in external space. In four dimensions these are scalars. Next we would have $2h^{1,2}$ vector fields and then another two vector fields in the last column. A 4-form field C_4 without self-duality condition would also come with another $h^{1,1} = h^{2,2}$ scalar fields that are associated with the harmonic $(2, 2)$ forms. We did not display this entry since only one set of $h^{1,1}$ scalars survives the self-duality condition. In the third and fourth column we imposed self-duality by multiplying these entries with a factor $1/2$. The overall massless content of type IIB supergravity on our CALABI-YAU manifold therefore reads

$$h^{1,2} \quad \text{vector multiplets}$$

$$\text{IIB:} \quad h^{1,1} + 1 \quad \text{hyper-multiplets}$$

$$1 \quad \text{gravitational multiplet.}$$

The result looks similar to the case of type IIA except that the numbers $h^{1,1}$ and $h^{2,1}$ got exchanged. A famous conjecture in mathematics states that for every CALABI-YAU space $\mathcal{X}$ one may find another $\tilde{\mathcal{X}}$ such that

$$h^{1,1}(\mathcal{X}) = h^{2,1}(\tilde{\mathcal{X}}), \quad h^{2,1}(\mathcal{X}) = h^{1,1}(\tilde{\mathcal{X}}).$$

A pair $(\mathcal{X}, \tilde{\mathcal{X}})$ of CALABI-YAU spaces with this property is known as a mirror pair. Our previous analysis shows that the massless content of type IIA theory on $\mathcal{X}$ agrees with that of type IIB theory on the mirror $\tilde{\mathcal{X}}$. It is believed that the statements extend to the entire string theory. We shall present a bit more evidence for such a symmetry between type IIA and IIB string theories in the following chapter.

17.2 KÄHLER Condition and GSO Projection

The KÄHLER condition and RICCI flatness gives CALABI-YAU spaces and a N=2 supersymmetric spectrum in 4-dimensional MINKOWSKI space, as we have just seen. On the world-sheet, consistency of the string background requires the existence of a super-VIRASORO algebra with $c = 15$. This allowed us to select physical states by imposing the super-VIRASORO constraints. But it was not sufficient for a successful construction of a tachyon-free spectrum. Recall from Chapter 10 that we also needed to perform a GSO projection. We have argued before that RICCI-flatness implies the existence of a super-VIRASORO algebra. We shall now argue that the KÄHLER condition is linked to the possibility of a consistent GSO projection.

It is useful to recall the form of the GSO projector from eqs. (10.13) and (10.14). In the NS-sector, for example, the right-moving GSO projector was given by

$$\Pi^{NS} = \frac{1}{2}(1 + \Gamma^{NS}_{\pm}) = \frac{1}{2}\left(1 - (-1)^{J_0}\right). \tag{17.5}$$

We have introduced the shorthand J_0 for the complicated looking exponent of (-1) in our construction of $\Gamma_{\pm}$ (see eq. (10.13)):

$$J_0 = i\sum_{j=0}^{4}\sum_{r=1/2}^{\infty}\left(b_{-r}^{2j}b_r^{2j+1} - b_{-r}^{2j+1}b_r^{2j}\right).$$

Here and in the following we shall omit the rotation of the modes b_r^0 by $-i$ for simplicity; i.e., we shall assume that all directions are space-like.

We are about to develop a new understanding of the GSO projection that rests on the curious observation that J_0 is actually the zero mode of the following U(1) current:

$$J(z) := \frac{i}{2}\sum_{j=0}^{4}\psi^{2j}(z)\psi^{2j+1}(z) = \sum_{n\in\mathbb{Z}}J_n z^{-n-1}. \tag{17.6}$$

As we shall see in a moment, $J(z)$ extends the N=1 super-VIRASORO algebra we had constructed in Chapter 8 to a larger algebraic structure that contains J, T and two fermionic fields $G^{\pm}$ that sum up to $\sqrt{2}G = G^+ + G^-$. Together, these form what is known as N=2 super-VIRASORO symmetry. Conversely, if the N=1 super-VIRASORO symmetry of some world-sheet model can be extended to N=2, then we can use the zero mode of J_0 for the construction of a GSO projector. In this sense, the N=2 super-VIRASORO algebra contains both the constraints that implement the physical state conditions and the GSO projection.

Let us first convince ourselves that our world-sheet model for type II superstrings in flat space contains an N=2 supersymmetric extension of the VIRASORO algebra.

To this end we rewrite the action (8.22) for strings moving in a $2n$-dimensional flat space with Euclidean metric in the form

$$S = \frac{1}{2\pi} \int d\tau d\sigma \left[\frac{2g_{ij}}{\alpha'} \partial_+ Z^i \partial_- Z^{\bar{j}} + g_{i\bar{j}} \, \psi^i_+ \partial_- \psi^{\bar{j}}_+ + g_{i\bar{j}} \, \psi^{\bar{j}}_- \partial_+ \psi^i_- \right], \quad (17.7)$$

where $Z^j, Z^{\bar{j}}$ and $\psi^j_\pm, \psi^{\bar{j}}_\pm$ with $j, \bar{j} = 1, \dots, n$ are complex combinations of our usual bosonic and fermionic fields X^α and $\psi^\alpha_\pm$:

$$Z^j = \frac{1}{\sqrt{2}} \left(X^{2j+1} + iX^{2j} \right), \quad Z^{\bar{j}} = \frac{1}{\sqrt{2}} \left(X^{2\bar{j}+1} - iX^{2\bar{j}} \right), \quad (17.8)$$

$$\psi^j_\pm = \frac{1}{\sqrt{2}} \left(\psi^{2j+1}_\pm + i\psi^{2j}_\pm \right), \quad \psi^{\bar{j}}_\pm = \frac{1}{\sqrt{2}} \left(\psi^{2\bar{j}+1}_\pm - i\psi^{2\bar{j}}_\pm \right). \quad (17.9)$$

After this rewriting it is readily checked that the action (17.7) is invariant under the following non-trivial supersymmetry transformations:

$$\delta Z^j = \epsilon_1 \sqrt{\frac{\alpha'}{2}} \, \psi^j_-, \quad \delta \psi^j_- = -\epsilon_2 \sqrt{\frac{\alpha'}{2}} \, \partial_- Z^j, \quad (17.10)$$

$$\delta Z^{\bar{j}} = \epsilon_2 \sqrt{\frac{2}{\alpha'}} \, \psi^{\bar{j}}_-, \quad \delta \psi^{\bar{j}}_- = -\epsilon_1 \sqrt{\frac{2}{\alpha'}} \, \partial_- Z^{\bar{j}}. \quad (17.11)$$

On the remaining fields $\psi^{\bar{j}}_+$ and ψ^j_+ the transformations δ are set to zero. Here, ϵ_1 and ϵ_2 are two independent constant Grassmann variables. In comparison with the supersymmetry transformations we discussed in Chapter 8 we split the first component ϵ_- of ϵ into $\epsilon_- = \epsilon_1 + \epsilon_2$ while setting the second component ϵ_+ to zero. Of course, there are also supersymmetry transformations with $\epsilon_+ = \tilde{\epsilon}_1 + \tilde{\epsilon}_2 \neq 0$ so that we find twice as many supersymmetry transformations as in our discussion of Chapter 8.

Under this extended supersymmetry, the stress energy tensor is part of a supermultiplet that contains two fermionic and one additional bosonic field. Explicitly, the fields of the supermultiplet are given by

$$T_{--}(\sigma^-) = -\frac{2}{\alpha'} g_{i\bar{j}} \, \partial_- Z^i \partial_- Z^{\bar{j}} - \frac{1}{2} g_{i\bar{j}} \, \psi^{\bar{j}}_- \partial_- \psi^i_- - \frac{1}{2} g_{i\bar{j}} \, \psi^i_- \partial_- \psi^{\bar{j}}_-, \quad (17.12)$$

$$G^+_-(\sigma^-) = \frac{2i}{\sqrt{\alpha'}} g_{i\bar{j}} \, \psi^{\bar{j}}_- \partial_- Z^i, \quad G^-_-(\sigma^-) = \frac{2i}{\sqrt{\alpha'}} g_{i\bar{j}} \, \psi^i_- \partial_- Z^{\bar{j}}. \quad (17.13)$$

The fields $G^\pm$ sum up to the fermionic field $\sqrt{2} G_- = G^+_- + G^-_-$ that we defined in Chapter 8. In addition to this, we define the U(1) current J as

$$J_-(\sigma^-) = -g_{i\bar{j}} \, \psi^{\bar{j}}_-(\sigma^-) \psi^i_-(\sigma_-). \quad (17.14)$$

It is easy to see that this definition of J agrees with the one we gave in eq. (17.6). The modes of $T, G^{\pm}$, and J generate an N=2 superconformal algebra with central charge $c = 3n$; i.e., they satisfy the following set of commutation relations:

$$[L_n, L_m] = (n - m)L_{n+m} + \frac{c}{12}n(n^2 - 1)\delta_{n,-m},$$

$$\left[L_n, G_m^{\pm}\right] = \left(\frac{n}{2} - m\right)G_{n+m}^{\pm},$$

$$[L_n, J_m] = -mJ_m,$$

$$\{G_n^+, G_m^-\} = 2L_{n+m} + (n - m)J_{n+m} + \frac{c}{3}\left(n^2 - \frac{1}{4}\right)\delta_{n,-m},$$

$$\left[J_n, G_m^{\pm}\right] = \pm G_{n+m}^{\pm}.$$

We have displayed the relations in the RAMOND sector where all the indices m, n assume integer values. The same relations also apply to the NEVEU-SCHWARZ sector only that the mode indices of $G^{\pm}$ take values in the set of half-integers.

We have now convinced ourselves that, at least in flat space, the existence of the U(1) current J whose zero mode appears in the GSO projector (17.5) follows from an enhanced world-sheet supersymmetry. This leaves us wondering under which circumstances such a supersymmetry enhancement can also appear for non-trivial compactifications. For curved backgrounds with a non-trivial metric $g_{\mu\nu} = g_{\mu\nu}(X)$, the action (8.22) can be extended as follows:

$$S = -\frac{1}{4\pi}\int d\tau d\sigma \left[\frac{g_{\mu\nu}}{\alpha'}\partial_a X^{\mu}\partial^a X^{\nu} + g_{\mu\nu}\bar{\psi}^{\mu}\rho^a D_a\psi^{\nu} + R_{\mu\nu\rho\sigma}(\bar{\psi}^{\mu}\psi^{\rho})(\bar{\psi}^{\nu}\psi^{\sigma})\right].$$
(17.15)

Here we used the same notations for fermions as in Chapter 8, R denotes the curvature tensor (16.22), and the covariant derivatives D_a are defined with the help of the CHRISTOFFEL symbols (16.20) as

$$D_a\psi^{\mu} = \partial_a\psi^{\mu} + \partial_a Z^{\nu}\Gamma^{\mu}_{\nu\rho}\psi^{\rho}.$$
(17.16)

The action (17.15) is invariant under N=(1,1) supersymmetry transformations.[1] These generalize the transformations (8.23) we considered in Chapter 8. In order to make things work, one has to add a non-linear term to the variation of the fermions.

In general, the action (17.15) does not possess N = (2,2) supersymmetry. In flat space, the enhancement was closely related to the split of space-time indices μ into holomorphic and anti-holomorphic ones. As we have seen in the previous chapter, such a split exists for all complex manifolds. But since the terms in the action are very sensitive to the metric, its CHRISTOFFEL symbols, and curvature

---

[1] The notation N = (1,1) refers to N=1 supersymmetry for left and for right movers.

tensor, it may not be too surprising that complex manifolds are not sufficient for supersymmetry enhancement. What one needs is that the metric is compatible with the complex structure, i.e. that the LEVI-CIVITA connections repects the split into holomorphic and anti-holomorphic indices, as is the case for KÄHLER manifolds. And indeed, it was shown by Zumino in [79] that the action (17.15) admits N = (2,2) supersymmetry if the metric $g_{\mu\nu}$ satisfies the KÄHLER condition, and hence the CHRISTOFFEL symbols and curvature tensor are given by eqs. (16.21) and (16.23), respectively. In this case, the stress energy tensor of the model is part of an N = (2,2) supermultiplet. If, in addition, g is RICCI flat, then the stress energy tensor T is also tranceless, and one may show the modes of T and its superpartners $G^{\pm}, J$ must satisfy the defining relations of an N=2 super-VIRASORO algebra. As we have argued before, this is sufficient to impose the physical state conditions and GSO projection. Hence, RICCI flat KÄHLER manifolds can give rise to consistent string backgrounds.

17.3 GEPNER Models: A Lightening Review

As we have discussed in the previous section, the description of strings in CALABI-YAU backgrounds requires an N = (2,2) superconformal field theory. The total central charge of the associated VIRASORO algebra must be $c = 15$ to satisfy the no-ghost condition. The 4-dimensional extended part of the model describing strings in flat MINKOWSKI space contributes $c_{\text{ext}} = 4 + 4 \times 1/2 = 6$. Hence, the compactification must contribute $c = c_{\text{int}} = 9$. The class of models we described in the previous section provide examples of superconformal field theories with the desired properties, but there are other ways to come up with N = (2,2) superconformal field theories of central charge $c = 9$.

We have seen before that we can pair real dimensions to build an N = (2,2) superconformal symmetry. This works in flat space, but also in some non-trivial backgrounds. The simplest example arises from the quotient SU(2)/U(1). In order to properly define this, we need to specify how the subgroup U(1) $\subset$ SU(2) acts on the numerator group SU(2). It turns out that the relevant action in our context is the adjoint one where elements $h \in$ U(1) act by conjugation $g \mapsto h^{-1}gh$ for $g \in$ SU(2). By an easy inspection we identify the resulting quotient space SU(2)/U(1) as a 2-dimensional disk. Its boundary consists of elements $g \in$ U(1) $\subset$ SU(2), which are fixed under the adjoint action of U(1). Since we know how to describe strings on the group manifold SU(2) (see Chapter 15), it may not come as a big surprise that we can also treat the quotient model with pretty much the same techniques we have seen in Chapter 15. The resulting theory is known as *parafermions*, and, as we know from the solution of problem 34, it contains generators of a VIRASORO algebra with central charge $c = c_k - 1 = (2k-2)/(k+2)$. See e.g. [18] for many more details about parafermionic conformal field theories.

But it is not quite the model we are after. As in flat space we need two additional MAJORANA fermions in order to be able to construct the $N = (2,2)$ super-VIRASORO algebra. In problem 29 we learned how to build the generators of an su(2) current algebra at level $k = 1$ from two fermions. Through the affine SUGAWARA construction these currents give rise to VIRASORO generators with central charge $c_1 = 1$. We can add these to the VIRASORO generators of the SU(2) WZNW model at level k to obtain an algebra with central charge $c = c_k + c_1$. Now we pass this model through the constructions outlined in the previous paragraph in order to obtain a quotient theory with VIRASORO central charge $c = c_k + c_1 - 1 = c_k$. With the additional fermionic sector, this family of models, known as $N = (2,2)$ *supersymmetric minimal models*, contains a full $N = (2,2)$ super-VIRASORO algebra with central charge:

$$c_k^{\text{SMM}} = \frac{3k}{k+2}.$$

Note that c_k is bounded by $c_k \leq 3$, i.e. by the value of the central charge that is contributed by a pair of flat directions in superstring theory.

In order to proceed with our conformal field theory construction of superstring "compactifications" we need a model with $N = (2,2)$ superconformal symmetry and central charge $c = c_{\text{int}} = 9$. A single supersymmetric minimal model is not sufficient to obtain such a large value of c. The idea of Doron Gepner in [23] was to multiply several minimal models with an appropriate set of levels k_v such that $c = \sum_v c_{k_v} = 9$. We will not discuss the details of this construction but would like to provide at least one example. It is obtained through multiplication of five supersymmetric minimal models with level $k = 3$. Since each factor contributes $c_3 = 9/5$ to the central charge, the product theory has central charge $c = 5 \times 9/5 = 9$, as required. The five factors we put together might remind us of the five complex coordinates we used for our construction of the quintic CALABI-YAU space in Chapter 16; see eq. (16.27). This seemingly superficial similarity goes much deeper. In fact, there exists very strong evidence that the GEPNER model we have just discussed describes the compactification of superstrings at a very special (small volume) point in the 102-dimensional moduli space of the quintic CALABI-YAU manifold.

Some of the key evidence for this relation is not that difficult to sketch; see, e.g., [30] for a review. Roughly speaking, it is indeed possible to detect the 102-dimensional parameter space right at the GEPNER point by studying deformations of the above GEPNER model. What this needs is a conformal field theory analogue of harmonic forms. Recall that our characterization of harmonic forms on CALABI-YAU spaces involves the differentials $\partial, \partial^\dagger$ and $\bar{\partial}, \bar{\partial}^\dagger$. In conformal field theory, the role of these differentials is played by the zero modes $G^\pm$ and $\bar{G}^\pm$. The zero modes of the left- and right-moving U(1) currents J and $\bar{J}$ in the

$N = (2,2)$ superconformal algebra are used to measure the form degree. A lightening review of GEPNER models and their relation with CALABI-YAU compactifications with many more details and references to the original literature can be found in [57].

Exercises

Problem 37. *(A) Determine the number of physical degrees of freedom for the following bosonic fields in a 6-dimensional space-time: Scalar, anti-symmetric tensor, graviton, vector.*
(B) There exist 2 types of massless $N = (1,1)$ supermultiplets in six dimensions: (i) The graviton multiplet consisting of a graviton, an anti-symmetric tensor, a scalar, and four vector fields; and (ii) the vector multiplet consisting of a vector and four scalar fields. What is the number of physical degrees of freedom for the corresponding supermultiplets (including fermions)?
(C) Type IIA supergravity compactified to a 4-dimensional K3 surface gives rise to a 6-dimensional $N = (1,1)$ supergravity theory. Determine the field content of this model.
HINTS: *(i) The HODGE diamond of the K3 surface is given by*

$$
\begin{array}{ccccc}
 & & 1 & & \\
 & 0 & & 0 & \\
1 & & 20 & & 1. \\
 & 0 & & 0 & \\
 & & 1 & &
\end{array}
\tag{17.17}
$$

(ii) The internal components of the 10-dimensional metric give rise to 58 scalars.

18

String Dualities

The last few chapters of this book deal with the vast subject of dualities in string and field theory. By the nature of its topic, our presentation will be significantly less self-contained and more descriptive. On the other hand, we hope to provide some introduction into a fascinating field of modern string theory that continues to expand.

All field and string theories depend on a certain number of parameters or couplings such as the string length l_s, the string coupling g_s, moduli of a compactification manifold, etc. In most cases, we cannot compute the dependence on the couplings exactly. Instead, perturbative techniques give us access to the neighborhood of some special points of the full parameter space, namely to those for which a perturbative description is known. It is difficult to determine the features of a model outside such weakly coupled regimes. On the other hand, it is certainly natural to wonder about the behavior of a theory at strong coupling. Many things can happen as we crank up one or more of the coupling constants. In particular, new fundamental degrees of freedom may show up, as it is the case, e.g., in QCD, where strongly coupled physics is best described in terms of hadrons, while quarks and gluons are the fundamental objects at weak coupling. Our main goal in this chapter is to explore type II string theories outside their perturbative regime. Our analysis will lead us to intriguing new relations, or dualities, between various perturbative expansions of string theory.

18.1 T-Duality and Mirror Symmetry

Before we start varying the string coupling g_s we shall briefly revisit one duality symmetry that we have seen before. To this end, let us go back to the case in which one coordinate of our string background is compactified to a circle of radius R.

Such a direction gives rise to the following contributions to the closed string mass operator:

$$M^2 = \frac{n^2}{R^2} + \frac{w^2 R^2}{\alpha'^2} - \frac{4}{\alpha'} + \frac{4}{\alpha'} \sum nN_n. \tag{18.1}$$

The first terms are associated with closed strings that move with some momentum and winding number n and w along the circular direction. This is followed by the usual tachyonic contributions and the excitation energies of the various string oscillations. We have observed before that this mass spectrum enjoys an interesting symmetry, namely the spectrum is invariant under a transformation $R \rightarrow \alpha'/R = R'$. In fact, if this transformation of the parameter R is accompanied by an exchange of the momentum and winding numbers n and w, the form of M^2 remains unchanged.

It is interesting to understand how branes and open strings experience this T-duality. Let us begin by looking at the operator M^2 for open string modes on a brane that satisfies DIRICHLET boundary conditions along the circle of radius R. It is not difficult to see that it has the form

$$M^2 = \frac{w^2 R^2}{\alpha'^2} - \frac{1}{\alpha'} + \frac{1}{\alpha'} \sum nN_n. \tag{18.2}$$

The expression is easy to interpret. Open strings with DIRICHLET boundary conditions must have both their ends at one fixed point on the circle. They can wind around the circle several times, but cannot carry any center of mass momentum.

If we impose NEUMANN boundary conditions along the circular direction instead of DIRICHLET ones, the physics changes a bit. In that case, the open string can carry center of mass momentum, but it cannot wind around the circle S^1. Hence, the mass operator takes a slightly different form:

$$M^2 = \frac{n^2}{R^2} - \frac{1}{\alpha'} + \frac{1}{\alpha'} \sum nN_n. \tag{18.3}$$

Suppose now we perform the usual T-duality transformation $R \rightarrow \alpha'/R$. Clearly, the spectrum (18.2) is no longer invariant under this transformation. Instead it gets mapped to the spectrum (18.3) of the mass operator for NEUMANN boundary conditions. In other words, T-duality remains an exact symmetry of a circle compactification even in the presence of D-branes. But it acts non-trivially on the choice of boundary conditions by exchanging DIRICHLET- and NEUMANN-type conditions.

The insights we gained from the previous discussion have some interesting implications for type II string theory. Recall that type II theories come in two species. Type IIA models contain Dp-branes that extend along an even number p of spacial directions. In type IIB theory, on the other hand, the number p must be odd. Let us suppose now that we have compactified, e.g., type IIA theory on a circle, and let us place some Dp-brane into this background. For definiteness we assume that the brane is located at some fixed position along the circle, i.e., we impose DIRICHLET boundary conditions along the circular direction. After T-duality, we obtain an equivalent theory provided we replace DIRICHLET boundary conditions along S^1 by NEUMANN boundary conditions. The new brane in the T-dual model extends along $p + 1$ spacial directions. Such a brane does not exist in type IIA theory. It rather belongs into the spectrum of a type IIB string compactification. Hence, we conclude that T-duality along a single circular coordinate maps type IIA to type IIB theory and vice versa. A symmetry between type IIA and IIB theories had briefly surfaced in the previous chapter when we discussed the massless spectrum of CALABI-YAU compactifications. The mirror symmetry we mentioned in that discussion may be considered as an extension of T-duality. Roughly speaking, mirror symmetry can be thought of as T-duality along three directions of a 6-dimensional compactification space.

Our discussion in this section has demonstrated how branes can act as useful probes to detect duality symmetries in string theory. We shall see much more of this as we proceed. Before approaching other examples of duality symmetries, however, let us briefly recall some of the central facts on branes and their properties.

18.2 Interlude on p-Branes in Type II Theories

The duality symmetries we are about to see will involve the string coupling g_s. It is therefore important for us to understand how string amplitudes depend on g_s. We shall discuss this first before we recall a few basic facts on D-branes.

A general amplitude in type II string theory can involve both open and closed string modes. The vertex operators that correspond to the states of string theory via the state-field correspondence are inserted at points on some RIEMANN surface with g handles and h boundary components; see Figure 18.1. As we shall argue now, such an amplitude is weighted with a factor

$$\mathcal{A}_{ST}^{g,h}(\phi_i, \psi_j) \sim g_s^{2g-2+h}, \tag{18.4}$$

where $i = 1, \ldots, N_c$, and $j = 1, \ldots, N_o$, label the external states of the scattering process. By definition, the string coupling g_s is a weight for the splitting and joining of closed strings. Since open strings are like half of a closed string, their splitting and joining comes with a weight $\sqrt{g_s}$. In order to find the weight of a general diagram, we need to count the number of pant diagrams that appear in the full amplitude. We shall not count this number directly, but rather determine how it changes as we increase the number N_c and N_o of closed and open string modes, the number g of handles, and the number h of boundary components.

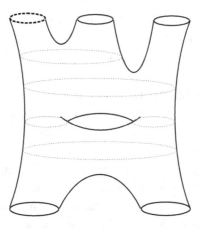

Figure 18.1 Pant decomposition of a string scattering diagram. Each handle adds two pairs of pants, while each boundary component and each external closed string state adds one pair of pants, as is shown in the upper left corner. The dashed line can be interpreted as a boundary component or an additional external closed string state.

Let us begin with N_c. If we add one additional closed string mode, we can join it with one of the original N_c closed string modes so that, after one joining process, we return to N_c insertions. Hence, adding one additional closed string state gives an extra factor g_s. Similarly, for an additional open string insertion, the amplitude gets multiplied by an extra factor $\sqrt{g_s}$. As for extra handles and boundary components, the effect of adding one may be analyzed as follows. Suppose that we add one additional handle. We can undo this operation by cutting the handle at some place. This will have the effect of adding two additional closed string states, one for each side of the cut. Hence, the processes with one additional handle are suppressed by a factor g_s^2. The same argument furnishes an extra factor g_s for each additional boundary component. Taking all this together, we have

$$\tilde{\mathcal{A}}_{\mathrm{ST}}^{g,h}(\phi_i, \psi_j) \sim g_s^{N_c + N_o/2 + 2g + h + a}.$$

Here, a is some constant that still needs to be determined. As we discussed in Chapters 4 and 7 already, we do not want the power of our loop counting parameter g_s to be raised with every single addition of an external state and therefore absorb one power of g_s into the normalization of the closed string states and similarly a factor $\sqrt{g_s}$ into each open string state. After such a "re-normalization" of the external states, a closed string tree-level amplitude with three external states is weighted with a factor $g_s g_s^{-3} = g_s^{-2}$. Similarly, an open string tree-level amplitude with three external open string states comes with a factor $\sqrt{g_s} g_s^{-3/2} = g_s^{-1}$. These

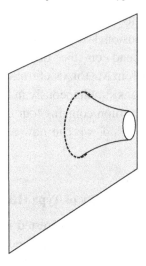

Figure 18.2 The leading contribution to the interaction of gravitons with a D-brane comes from world-sheets with a single boundary component and one closed string insertion.

factors are reproduced by our formula (18.4), and they determine the constant a in the previous formula to be $a = -2$.

Let us now apply these results to determine how the mass of a D-brane depends on the string coupling constant g_s. Note that the mass determines the strength of the coupling of the D-brane to the graviton. As shown in Figure 18.2, the interaction graviton and brane receives its leading contribution from an amplitude with a single boundary component and one closed string mode insertion. In our notations above, this corresponds to $h = 1$ and $g = 0$. We conclude that the tension of a Dp-brane is given by

$$T_{\mathrm{D}p} \sim \frac{1}{g_s}\frac{1}{l_s^{p+1}}. \tag{18.5}$$

The exponent of g_s is given by $2g + h - 2 = -1$. The dependence on the string length l_s is obvious for dimensional reasons. As we discussed in great detail in Chapters 7 and 11, excitations of Dp-branes are described by open strings. In type II theories, their low-energy effective action is obtained by dimensional reduction of an N=1 supersymmetric YANG-MILLS theory from 10 to $p + 1$ dimensions,

$$S[A, \psi] = \frac{1}{g_s l_s^{p-3}} \int \mathrm{d}^{p+1}x \, \mathrm{tr}\left(F_{ij}F^{ij} + D_i\phi_n D^i\phi^n + [\phi_n, \phi_m][\phi^n, \phi^m]\right)$$
$$+\text{fermionic terms}, \tag{18.6}$$

where the covariant derivatives D_i are given by $D_i = \partial_i + A_i$, as usual, and the trace is taken in the space of $N \times N$ matrices. In 10 dimensions, the MAJORANA-WEYL fermion contains 16 components. These provide fermionic fields in the field theory on the brane's $(p + 1)$-dimensional world-volume. The number and nature of fermions on the brane depend on the dimension of the D-brane. When $p = 3$, for example, we obtain four MAJORANA fermions, because we cannot impose MAJORANA and WEYL conditions simultaneously in four dimensions. Furthermore, a 4-dimensional MAJORANA fermion contains four components. With these rather simple facts on Dp-branes in mind, we can now return to the main goal of this chapter.

18.3 Self-Duality of Type IIB Theory

For the rest of the chapter, we shall be interested in understanding the behavior of type II string theories at strong coupling g_s. Finding an efficient description of strongly coupled string theory might seem to be a very daunting task. In the context of QCD, for example, the analogous question would be to derive the physics of hadrons from perturbative QCD. Other examples exist, however, mostly in the context of 2-dimensional quantum field theory, where such issues have been addressed very successfully. As we shall see now, the structure of strongly coupled type II theories is also more accessible than one might naively expect.

Once more, we shall use branes to probe properties of type II theories as we crank up the coupling g_s. Let us begin with type IIB theory and ask what happens to the D1-branes at strong coupling. From the mass formula (18.5) we conclude that the tension of a D1-brane decreases with growing coupling g_s. For small coupling, it is certainly much larger than the tension

$$T_{F1} = \frac{1}{l_s^2}$$

of the fundamental string. But while g_s increases to very large values, the tension of the D1-brane can become smaller than that of the fundamental string. In that case, the D1-brane turns into the fundamental object of the theory. For a closer look at the behavior of D1-branes in the type IIB theory we shall study its dynamics. Recall that excitations of a single D1-brane are described by the low-energy effective action

$$S_{D1} \sim \frac{l_s^2}{g_s} \int d^2x \left(F_{ij}F^{ij} + D_i\phi_n D^i\phi^n \right) + \cdots$$

$$\sim \frac{l_s^2}{g_s} \int d^2x\, \partial_i\phi_n \partial^i\phi^n + \cdots ,$$

where the index n runs from $n = 2$ to $n = 9$. Following the general recipe we discussed in Chapter 11 this action may be obtained by dimensional reduction from N=1 supersymmetric Maxwell theory in 10 dimensions. Since we are discussing a single D-brane, all fields of the theory take values in scalars so that the commutator terms vanish. In passing from the first to the second line we have dropped the gauge fields simply because they do not possess any physical degrees of freedom on the 2-dimensional world-volume of the D1-brane. As before, the $\cdots$ stand for fermionic contributions. Let us sketch, at least qualitatively, what the fermionic sector of the model looks like. Before the dimensional reduction, the MAJORANA-WEYL fermions of the model possess 16 components. After the reduction to the 2-dimensional world-volume of the D1-brane, these decompose into eight MAJORANA-WEYL fermions. Hence, the dynamics of a D1-brane in a flat 10-dimensional background is described by the theory of eight free bosons and the same number of MAJORANA-WEYL fermions. This is a field theory we are quite familiar with. In fact, it is used to model the physical degrees of freedom of a fundamental string in the 10-dimensional MINKOWSKI space. Consequently, a D1-brane of type IIB theory behaves like a fundamental string if we replace $l_s^2 \rightarrow g_s l_s^2$. We now conjecture that type IIB string theory is self-dual under the following replacements of its coupling constants:

$$(g_s, l_s) \quad \leftrightarrow \quad (g_s' = g_s^{-1}, l_s' = \sqrt{g_s} l_s) . \tag{18.7}$$

The replacement was designed such that it maps the tension of the D1-brane to that of the fundamental string and vice versa:

$$T_{D1}' \;=\; \frac{1}{g_s' l_s'^2} \;=\; \frac{g_s}{g_s l_s^2} \;=\; \frac{1}{l_s^2} \;=\; T_{F1} . \tag{18.8}$$

This test alone would hardly be convincing if it was not for the other branes of type IIB superstring theory. Recall that type IIB theory also contains D3- and D5-branes. For the tension of the D3-brane we find

$$T_{D3}' \;=\; \frac{1}{g_s' l_s'^4} \;=\; \frac{g_s}{g_s^2 l_s^4} \;=\; \frac{1}{g_s l_s^4} \;=\; T_{D3} . \tag{18.9}$$

The computation indicates that the D3-brane in strongly coupled IIB theory behaves like a D3-brane in the weakly coupled model. The dynamics of D3-branes in the type IIB theory is described by a YANG-MILLS theory with 16 supercharges, i.e. by N=4 supersymmetric YANG-MILLS theory. According to the general rules, the coefficient of this theory is given by $1/g_s = 1/g_{YM}^2$. Hence, large values of g_s correspond to strongly coupled gauge theory. In order for the conjecture on the self-duality of type IIB theory to be correct, strongly coupled N=4 super YANG-MILLS theory should behave just like the weakly coupled model. This field theoretic conjecture was actually formulated a few decades ago in the work of Montonen and Olive [45]. In fact, much evidence has been assembled for the self-duality of N=4 super-YANG-MILLS theory under the transformation $g_{YM} \leftrightarrow g_{YM}^{-1}$. Therefore, the conjecture on the self-duality of type IIB theory ties in beautifully with a very non-trivial statement of 4-dimensional quantum field theory.

It remains to look at the D5-brane of type IIB theory. With our usual replacement rules we find that

$$T_{D5} = \frac{1}{g_s'(l_s')^6} = \frac{g_s}{g_s^3 l_s^6} = \frac{1}{g_s^2 l_s^6}. \tag{18.10}$$

Here we are facing a new problem. None of the D-branes in type IIB theory possesses a tension that goes like g_s^{-2}. At this point let us recall a brief remark we made in Chapter 11 when we discussed classical solutions of type II supergravities. There, we mentioned the existence of solitonic solutions that source the KALB-RAMOND field and its dual rather than the RR-fields of the theory. One of these solutions was identified as the fundamental string, while the other is known as the NS5-brane. We did not investigate such NS5-branes any further. But if we believe in the self-duality of type IIB theory, then the tension we found in the previous computation must be that of an NS5-brane:

$$T_{NS5} = \frac{1}{g_s^2 l_s^6}.$$

From the dependence on the string coupling g_s we conclude that the description of NS5-branes must be very different from that of Dp-branes. Note that the factor g_s^{-2} is characteristic for closed string tree-level amplitudes. Hence, while the effective action of Dp-branes involves open string modes, that of NS5-branes must arise from closed string excitations. This closed string theory describing NS5-branes has been explored; see [40]. It involves the SU(2) WESS-ZUMINO-NOVIKOV-WITTEN model we have studied in detail in Chapter 15.

18.4 Strongly Coupled IIA Theory

In the last section we studied the behavior of strongly coupled IIB theory. We collected considerable support for the claim that type IIB string theory looks exactly the same at strong and weak coupling. It is now very natural to wonder about the strong coupling limit of type IIA theory. Clearly, the story must be somewhat different. Type IIA theory contains D0-, D2-, and D4-branes in addition to the fundamental string and the NS5-brane. Hence, there are no two objects with the same dimensionality.

The key to strongly coupled IIA theory comes from the study of D0-branes. Recall that a single D0-brane has mass $M_{D0} \sim 1/g_s l_s$. We shall consider a stack of N of such D0-branes. These branes interact with each other, and one may wonder whether they can form some kind of bound states. We will now argue that this is the case. In fact, we shall describe evidence for the formation of so-called marginal bound states. The latter possess mass $M_N \sim N/g_s l_s$. Vanishing of their binding energy can be traced back to the supersymmetry of the configuration. On the other hand, one may show that the number of states with mass $M_N = N/g_s l_s$ is independent of N. For a single D0-brane the massless excitations are acted upon by the superalgebra. As we recalled before, there are 16 supercharges. While half of them decouple in massless representations, the remaining eight can be used to build four creation operators so that we obtain 2^4 ground states. Let us now increase the number of D0-branes. If N D0-branes would not form any bound state, the degeneracy of ground states would increase rapidly with the number N. But this is not what happens. Instead, the degeneracy of ground states remains at the 2^4. Explaining the source of such insights requires some discussion.

When D-branes are close together, their dynamics is well approximated by the dynamics of massless open string modes that stretch between them. For our N D0-branes, the corresponding low-energy effective action involves matrix valued fields that depend on a single variable x_0. The precise form of the action is again obtained from the action of N=1 super YANG-MILLS theory by dimensional reduction from 10 dimensions to a single time variable x_0. Because the dynamical variables are $N \times N$ matrices, such a model is referred to as a *matrix model*. From the classical action one can work out the HAMILTON function and the POISSON brackets. Canonical quantization provides us with a HAMILTON operator. Since all these steps are straightforward, we shall not describe the details any further.

The resulting HAMILTON operator has been studied very extensively. From the physics it describes and our comments above we conclude that it must be of the form

$$H_{\text{U}(N)} = \frac{g_s l_s}{2N}(P_{\text{c.m.}})^2 + H_{\text{SU}(N)}. \qquad (18.11)$$

On the right-hand side we have simply separated the kinetic energy for the motion of the center of mass from the term that describes the binding energy. The coefficient of the first term contains the mass parameter $M_N \sim N/g_s l_s$. Supersymmetry implies that the eigenvalues of the second term $H_{\text{SU}(N)}$ are bounded from below by zero. There exist eigenstates with zero eigenvalue, and so N D0-branes do not release any energy while they form their bound states. The most non-trivial insight into $H_{\text{SU}(N)}$ concerns the number of states with zero energy. As found by Sethi and Stern [61], and then proven by Hoppe et al. (see [36] for references), the Hamiltonian $H_{\text{SU}(N)}$ for the relative motion of N D0-branes possesses 2^4 normalizable ground state with zero energy. This degeneracy is independent of N. Note that all these statements on $H_{\text{U}(N)}$ can be formulated and proven rigorously.

One may wonder now what these findings imply for our analysis of strongly coupled type IIA theory. In order to understand the connection, let us briefly recall some facts on the KALUZA-KLEIN reduction of a massless scalar field on $\mathbb{R}^{1,D-1} \times S_R^1$. Such a field obeys the usual KLEIN-GORDON equation

$$\left(\frac{\partial^2}{\partial x_0^2} - \sum_{i=1}^{D-1} \frac{\partial^2}{\partial x_i^2} - \frac{\partial^2}{\partial \theta^2} \right) \phi(x, \theta) = 0, \tag{18.12}$$

where $\theta \in [0, 2\pi R[$ is the coordinate of S_R^1. Using the Ansatz $\phi(x, \theta) = \sum \phi_N(x) \exp(iN\theta/R)$ we can split this equation into an infinite number of equations for the FOURIER coefficients $\phi_N(x)$:

$$\left(\frac{\partial^2}{\partial x_0^2} - \sum_{i=1}^{D-1} \frac{\partial^2}{\partial x_i^2} - \frac{N^2}{R^2} \right) \phi_N(x) = 0. \tag{18.13}$$

So, if we interpret the theory of a single scalar field on $\mathbb{R}^{1,D-1} \times S_R^1$ from the perspective of an observer in $\mathbb{R}^{1,D-1}$, we see an infinite tower of scalar fields with linear mass spectrum $M_N = N/R$.

Such a simple KALUZA-KLEIN mass spectrum is exactly what we obtained from the marginal bound states of D0-branes in type IIA theory. In that case, the radius parameter is given by $R = g_s l_s$. As long as the string coupling g_s is small, the D0-branes are extremely heavy, and we do not see much of the marginal bound states. But as we increase the coupling, the bound states of D0-branes become lighter. In view of our discussion of KALUZA-KLEIN spectra we now interpret this behavior in terms of an additional 11th dimension that is rolled up on a circle of radius $R = g_s l_s$. For small values of g_s, the circle has a small radius so that the KALUZA-KLEIN modes possess a large mass. But as we increase the coupling g_s, the size of the circle grows and the mass of the KALUZA-KLEIN excitations comes down. These observations lead to the idea that type IIA string theory with couplings (g_s, l_s) can be obtained from an 11-dimensional theory, called *M-theory*, by compactification on a circle with radius $R \sim g_s l_s$.

Not much is known about M-theory. In particular, we do not have any description that would allow us to determine its spectrum and scattering amplitudes in the same way as this was done for 10-dimensional string theories. One thing we can say, however, concerns the low-energy limit. In fact, the low-energy behavior of M-theory must be described by an 11-dimensional gravitational theory with 32 supercharges. There is only one such theory, the unique 11-dimensional N=1 supergravity. Its action reads

$$S_{11D} \sim \frac{1}{G_{11}} \int d^{11}x \left[\sqrt{-g} \left(R + \frac{1}{48} F^2 \right) + \frac{1}{6} A \wedge F \wedge F \right], \tag{18.14}$$

where $F = dA$ is a 4-form field strength. As in the case of 10-dimensional supergravities, this 11-dimensional model possesses solitonic solutions. Since the only form potential of the model is the 3-form A, there should exist 3-dimensional solitons that can source this form field. In addition, one can also construct some dual 6-dimensional solitonic objects. These two types of solitonic solutions are known as the M2- and M5-brane, respectively.

Somehow, the M2- and M5-brane of M-theory must be related to the branes in type IIA supergravity after M-theory gets compactified on a circle of radius $R = g_s l_s$. Let us discuss these relations for M2-branes first. There are two things we can do with an M2-brane in M-theory on $\mathbb{R}^{1,9} \times S^1$. One possibility is to place it at some point of the circle and have its two spacial dimensions extend along the uncompactified MINKOWSKI space. If our M-theory compactification is indeed related to type IIA string theory, then we should obtain the D2-brane. There is not much we can do to test such a relation. But we can use it to determine the tension $T_{M2} = T_{D2}$ of the M2-brane. The result is shown in the first line of the table below. A qualitatively different situation arises if we let the M2-brane wrap around the circle S^1 of radius $R = g_s l_s$. An observer in $\mathbb{R}^{1,9}$ will now see an object that extends along a single spacial direction with tension (mass density) given by $T = T_{M2}R$. This tension is determined in the second line of the following table, and it is found to agree with the tension of a fundamental string in type IIA superstring theory.

$$\text{M-theory} \qquad \text{type} \quad \text{IIA theory on} \quad \mathbb{R}^{1,9} \times S_R^1 \, ,$$

$$\text{M2} \quad \rightarrow \quad \text{D2} \quad \text{with tension} \quad T_{D2} = 1/g_s l_s^3 =: T_{M2} \, ,$$

$$\searrow \quad \text{F1} \quad \text{with tension} \quad T_{M2}R = 1/l_s^2 = T_{F1} \, ,$$

$$\text{M5} \quad \rightarrow \quad \text{NS5} \quad \text{with tension} \quad T_{NS5} = 1/g_s^2 l_s^6 =: T_{M5} \, ,$$

$$\searrow \quad \text{D4} \quad \text{with tension} \quad T_{M5}R = 1/g_s l_s^5 = T_{D4} \, .$$

A similar analysis applies to the M5-brane in an M-theory compactification on $\mathbb{R}^{1,9} \times S^1$. When the M5-brane is placed at a fixed point of the circle, we obtain the NS5-brane of type IIA theory. D4-branes, on the other hand, arise from M5-branes that wrap the circle S^1. This simple match between brane spectra supports the suggested relation between M- and type IIA theory.

The main results of this chapter, along with some others we did not discuss, are all summarized in Figure 18.3. It visualizes the various relations between type II theories and M-theory we have outlined above, including the self-duality of type IIB theory, its T-duality with type IIA models, and the duality between type IIA and M-theory. Through the duality symmetries, all these models appear as different aspects of a single underlying model. Similar studies of type I and heterotic string compactifications have been performed. The results are summarized on the right-hand side of Figure 18.3. When read clockwise from the top, the figure describes

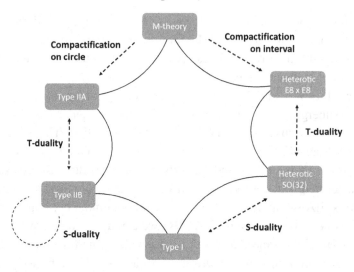

Figure 18.3 Dualities between M-theory and the different string theories.

that heterotic $E_8 \times E_8$ string theory may be obtained by compactifying M-theory on an interval $S^1/\mathbb{Z}_2$. Furthermore, the two consistent heterotic models with gauge groups $E_8 \times E_8$ and SO(32) are T-dual to each other. Finally, the strongly coupled SO(32) model may be reproduced by type I string theory. What we have described in previous chapters about heterotic and type I theories is not sufficient to verify this part of Figure 18.3. But in the light of our analysis for the left-hand side of the figure, the extensions to heterotic and type I models may seem a little less mysterious.

19

Gauge/String Dualities

At the end of the last chapter we have seen that strongly coupled theories may have unexpected descriptions. These can look very different from the weakly coupled formulation of the model. In the case of type IIA/M-theory duality, for example, it was suggested that we describe strongly coupled type IIA superstring theory in terms of an 11-dimensional theory. The only candidates for a fundamental description of the latter were the M2- and M5-branes of 11-dimensional supergravity. Admittedly, the duality between the two theories remained rather speculative, mostly because it is not possible to give a proper construction of M-theory. In this section we shall see another example of a duality that involves two theories that are reasonably well defined, yet very different in nature. It relates $(p + 1)$-dimensional gauge theories with 10-dimensional type II string compactifications. As in the example of type IIA/M-theory, both theories live in a space-time of different dimension, and their fundamental degrees of freedom are quite distinct. Special attention will be paid to the simplest example of such a duality between 4-dimensional N=4 Super YANG-MILLS (SYM) theory and closed strings on $AdS_5 \times S^5$. This will enable us to outline the formulation and the use of such dualities.

19.1 'T HOOFT's Analysis of the Large N_c Limit

The idea of a duality between gauge and string theories is not new. Even though the first concrete examples of gauge/string dualities were formulated only at the end of the 1990s, it had long been suspected that large N_c gauge theories may be re-written in terms of closed strings. We shall describe some of these arguments in this section.

We shall begin by reviewing an argument of G.'t Hooft [66] that shows that computations in gauge theory can be rearranged such that they look like an expansion in some model of closed strings. Suppose we are given a pure

YANG-MILLS theory with gauge group $U(N_c)$. Schematically, the action of this theory takes the form

$$S \sim \frac{1}{g_{\mathrm{YM}}^2} \int d^D x \, \mathrm{tr} \left(\partial_\mu A_\nu \partial^\mu A^\nu - \partial_\mu A_\nu \partial^\nu A^\mu + 2 \partial_\nu A_\mu [A^\mu, A^\nu] + \cdots \right).$$

Here we have omitted terms of fourth order since they are not relevant for what we are about to discuss. From the action we may read off the FEYNMAN rules. Let us recall that the components A_μ are actually $N_c \times N_c$ matrices, i.e., A_μ is a shorthand for N_c^2 fields A_μ^{ab} with $a, b = 1, \ldots, N_c$. Therefore, the theory contains N_c^2 propagators that are labeled by the two colors a and b. It is customary to picture these propagators as double lines where each of the two lines carries a color charge. Such double lines are then joined together through vertices such that the color charge is conserved. For the propagator we read off

$$\rule[2mm]{2.2cm}{0.4pt} \rule[0.5mm]{2.2cm}{0.4pt} \;\; \sim \; g_{\mathrm{YM}}^2$$

while for the 3-vertex one finds

$$\sim \; \frac{1}{g_{\mathrm{YM}}^2} \, .$$

We would now like to evaluate the partition function of the theory by summing up all FEYNMAN diagrams without external legs. As we have just reviewed, we can picture individual contributions as double line networks Γ in which each line carries a definite color charge. Our interest is not so much in evaluating contributions but rather to understand their dependence on the YANG-MILLS coupling g_{YM} and the number N_c of colors.

This turns out to be relatively easy. Note that each propagator comes with a factor g_{YM}^2 while each vertex contributes g_{YM}^{-2}. Therefore, the partition function takes the form

$$Z^{\mathrm{YM}} \sim \sum_\Gamma g_{\mathrm{YM}}^{2E-2V} N_c^L f(\Gamma), \tag{19.1}$$

where $E = E(\Gamma)$ is the total number of propagators (edges) in the graph Γ and $V = V(\Gamma)$ devotes the number of vertices. The factor N_c^L arises from a summation over the color indices that run through the L loops of the graph Γ. Our expression

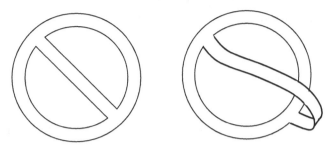

Figure 19.1 Two examples of double line networks that contribute to the partition function of pure gauge theory. While the left one can be drawn on a plane/sphere without intersections, we need a torus for an intersection-free drawing of the right one.

for Z^{YM} displays all dependence on g_{YM} and N_c. The remaining functions $f(\Gamma)$ are independent of these parameters. For complicated double line networks Γ, the functions $f(\Gamma)$ are very difficult to compute. But the following argument does not need any more detailed knowledge about these functions.

To each double line network we can associate a number (index) g through $2g = E - L - V + 2$. All double line networks with index g can be drawn on a 2-dimensional RIEMANN surface of genus g. Rather than reviewing the general argument we shall discuss two simple examples. To begin, consider the diagram shown on the left side of Figure 19.1. It is built from $E = 3$ propagators, $V = 2$ vertices, and it contains $L = 3$ closed color loops. Hence, the associated index is $2g = 3 - 3 - 2 + 2 = 0$. And, indeed, it is possible to draw this diagram onto the RIEMANN sphere, simply by filling in all the closed loops. The second example is depicted on the right side of Figure 19.1 It contains $E = 3$ propagators, $V = 2$ vertices, and a single color loop, $L = 1$. In this case the overall index is $2g = 3 - 2 - 1 + 2 = 2$. This agrees with the fact that the second double line network in Figure 19.1 can be drawn onto a 2-dimensional torus.

After this preparation we now rewrite the expression (19.1). The idea is to re-group the summation over all double line networks such that we sum over all double line networks Γ_g of given genus g first and then add up all the contributions by summing over g:

$$Z^{\text{YM}} \sim \sum_{\Gamma} g_{\text{YM}}^{2E-2V} N_c^{E-V} N_c^{L+V-E} f(\Gamma) = \sum_{g=0}^{\infty} N_c^{2-2g} \sum_{\Gamma_g} \lambda^{E-V} f(\Gamma_g). \quad (19.2)$$

Here we also introduced the so-called 'T HOOFT coupling, $\lambda = g_{\text{YM}}^2 N_c$. In this way, the partition function is now organized in a sum over the genus g, just as the loop expansion of a closed string theory. Recall that a genus g amplitude of closed string theory comes weighted by a factor g_s^{2g-2}. So, if we identify $g_s \sim 1/N_c$, we can write the partition function as

$$Z^{YM} \sim \sum_{g=0}^{\infty} g_s^{2g-2} F_g(\mathcal{X}_\lambda) \quad \text{with} \quad F_g(\mathcal{X}_\lambda) = \sum_{\Gamma_g} \lambda^{E-V} f(\Gamma_g). \tag{19.3}$$

In this way, the partition function of YANG-MILLS theory appears as if it had come from a theory of closed strings in some unknown background $\mathcal{X}_\lambda$ that depends on the strength of the 'T HOOFT coupling λ.

Before we close our discussion of the large color limit of gauge theories we would like to mention one possible extension. Note that the theory we discussed contains only matrix valued fields A_μ. The same is true for the gauge theories we have realized as a low-energy effective field theory on a stack of parallel D-branes in MINKOWSKI space. On the other hand, nature also contains quarks, i.e. objects carrying a single color charge. Their propagators are represented by single lines. In a theory of quarks and gauge fields, the FEYNMAN graphs take the form of networks with both double and single lines. Each single line propagator contributes a factor g_{YM}^2, just as for the gluon propagators. Similarly, vertices give g_{YM}^{-2}. The only new feature comes from the closed loops since some of them are no longer surrounded by a colored line. If we continue to denote the total number of loops in the network by L and introduce L_s to count the number of single line loops, we obtain

$$Z^{YMQ} \sim \sum_{\Gamma} g_{YM}^{2E-2V} N_c^{L-L_s} f(\Gamma) = \sum_{g,h} g_s^{2g+b-2} F_{g,h}(\mathcal{X}_\lambda). \tag{19.4}$$

Here, we have introduced the sum over the genus g of the graph Γ, as before, and we renamed $L_s = b$. Hence, the partition function of a gauge theory containing fields with a single color charge takes the form of an expansion in open string theory. The number L_s of single line loops corresponds to the number b of boundary components on the world-sheet.

While 'T HOOFT managed to re-write gauge theory quantities in a form that resembles an expansion of some string theory, it does not give us any hint which string background $\mathcal{X}_\lambda$ we should choose, how this choice depends on the field content, and the coupling constants of the gauge theory. But with the insight we gained in the course of these chapters, and in particular into the physics of D-branes, we can do much better. Before we turn to some concrete examples, we invite the reader to re-read the final section of chapter 1. There we discussed the dual role open and closed strings assume when we analyze the interaction between D-branes, and we commented on a limiting case in which the duality involves gauge fields instead of open strings. It is very tempting to connect this discussion to the large N_c limit of gauge theory by conjecturing that the string background $\mathcal{X}_\lambda$, which was left undetermined by 'T HOOFT's analysis, must look like the geometry in the vicinity of Dp-branes. This is the content of the celebrated AdS/CFT correspondence.

19.2 A Dual Description of N=4 SYM

In the introduction we argued that modern string theory should be able to produce fascinating novel relations between gauge and string theory. But the picture was a bit too general to fully appreciate the powerful implications of our discussion. In order to be more specific, let us focus on the most studied example. It arises from a stack of N_c parallel D3-branes that are placed in a flat 10-dimensional type IIB superstring background [42]. We have spelled out the corresponding solution of type IIB supergravity in Chapter 11. Here we shall recall only the form of the metric

$$ds^2 = f_3^{-\frac{1}{2}} dx^2 + f_3^{\frac{1}{2}} \left(dr^2 + r^2 d\Omega_5 \right) \quad \text{where} \quad f_3 = 1 + \frac{G_{10} T_{D3} N_c}{4\Omega_5} \frac{1}{r^4}. \quad (19.5)$$

We have written the tension of the whole stack as a product of the tension T_{D3} for a single D3-brane and the number N_c of branes. In addition the solution came with a non-trivial RR 4-form C_4. Note that for the special case of $p = 3$ the dilaton field is constant.

According to our general discussion, low-energy excitations on such a brane configuration are described by some $(3 + 1)$-dimensional gauge theory. The theory in question is obtained by dimensional reduction from N=1 Yang Mills theory in 10 dimensions. We have discussed this theory at the end of Chapter 11 and in the previous chapter. Since N=1 supersymmetry in 10 dimensions involves 16 supercharges, the 4-dimensional reduced theory must have N=4 supersymmetry. In addition to the U(N_c) gauge bosons, the model possesses six matrix valued scalar fields, one for each direction transverse to the branes, and a bunch of fermions; see Chapter 11. Admittedly, except for being 4-dimensional, this is not the most realistic model of our world. Not only does it possess the wrong matter content, it also is an example of a conformal field theory (CFT) – i.e., it looks exactly the same on all length scales, in sharp contrast to, e.g., QCD. In particular, the N=4 super YANG-MILLS quantum theory has no confining phase. But for the moment we intend only to explain some general ideas, and so we defer such concerns to a later stage of our discussion.

Gauge/string duality claims that, in the limit of large number N_c of colors (see below for a more detailed discussion), N=4 super YANG-MILLS theory is dual to a theory of closed strings that propagate in the curved near-horizon geometry of our stack of D3-branes. Let us investigate what this geometry looks like. For small distances r from the D3-branes, we can write the function f_3 as

$$f_3 = 1 + \frac{G_{10} T_3 N_c}{4\Omega_5} \frac{1}{r^4} \sim \frac{R^4}{r^4}, \tag{19.6}$$

where we have introduced the dimensionful parameter

$$R^4 = \frac{G_{10} T_3 N_c}{4\Omega_5} \sim \frac{g_s^2 l_s^8 N_c}{g_s^2 l_s^4} = g_s N_c l_s^4. \tag{19.7}$$

In this short computation we inserted the 10-dimensional gravitational constant $G_{10} \sim g_s^2 l_s^8$. Its inverse G_{10}^{-1} appears in front of the supergravity action and hence must be proportional to $g_s^{2g+b-2} \sim g_s^{-2}$ for tree-level amplitudes ($g = 0$) in closed string theory ($b = 0$). The dependence of G_{10} on l_s is fixed such that G_{10} has the correct dimension. In addition, we recall the expression $T_{D3} = 1/g_s l_s^4$ from the previous chapter. Plugging our result for f_3 back into the metric, we obtain

$$ds^2 = \frac{r^2}{R^2} dx^2 + \frac{R^2}{r^2} dr^2 + R^2 d\Omega_5^2. \tag{19.8}$$

We see that ds^2 splits into a sum of two terms. $R^2 d\Omega_5^2$ is simply the metric of a compact 5-sphere S^5 of radius R. The other terms correspond to a 5-dimensional anti-de Sitter (AdS) space AdS_5. A 5-sphere S^5 consists of all points in 6-dimensional Euclidean space that possess the same distance R from the origin. The anti-de Sitter space AdS_5 can be constructed in the same way, only that the 6-dimensional Euclidean space is now replaced by a space with two time-like and four space-like coordinates, and the defining equation for AdS_5 becomes

$$y_0^2 + y_5^2 - y_1^2 - y_2^2 - y_3^2 - y_4^2 = R^2 . \quad (AdS_5) \tag{19.9}$$

One can parametrize the 5-dimensional space of solutions to this equation in terms of a 4-vector x and a radial coordinate r as follows:

$$y_0 = Rt/r, \qquad y_5 = \frac{r}{2}\left(1 + r^{-2}(R^2 + x^2)\right), \tag{19.10}$$

$$y_i = Rx_i/r, \qquad y_4 = \frac{r}{2}\left(1 - r^{-2}(R^2 - x^2)\right), \tag{19.11}$$

where $i = 1, 2, 3$ and $x^2 = -x_0^2 + x_1^2 + x_2^2 + x_3^2$ as before. If this parametrization is inserted into the expression for the metric on the 6-dimensional embedding space

$$ds^2 = -dy_0^2 - dy_5^2 + dy_1^2 + dy_2^2 + dy_3^2 + dy_4^2, \tag{19.12}$$

we obtain those contributions in eq. (19.8) that are not associated with the 5-sphere. Let us point out in passing that the near horizon geometry of our stack of D3-branes possesses more symmetry than one would have naively expected. In fact, points on a 5-sphere are acted upon by elements of an SO(6). Similarly, the *AdS₅* geometry described through eqs. (19.9) and (19.12) is obviously left invariant by elements of the non-compact group SO(2,4). This observation will become very important later.

The assertion of the duality is that computations in N=4 super YANG-MILLS theory and in string theory on *AdS₅* × *S⁵* give identical results! Of course, gauge physicists and string theorists need to use an extensive dictionary in order to compare the outcome of their respective computations. We shall discuss a few entries of this dictionary as we proceed in order to give an idea of how this works.

To begin, let us talk about the parameters in the two theories. On the gauge theory side, there are two of them: the number N_c of colors and the YANG-MILLS coupling g_{YM}. In the context of large N_c limits, it is more appropriate to work with the 'T HOOFT coupling $\lambda = g_{YM} N_c^2$ instead of g_{YM}. On the string theory side, we have the string coupling g_s and the radius R/l_s of the *AdS₅* space measured in units of the string length l_s. We are prepared now to state the first entry in the AdS/CFT dictionary, which relates the two sets of parameters. Solving the equation (19.7) for the number N_c of colors, we find

$$N_c = \frac{R^4}{l_s^4} g_s^{-1}. \qquad (19.13)$$

So, up to the multiplication with the dimensionless quantity, the string loop parameter g_s is proportional to $1/N_c$, just as in 'T HOOFT's analysis. In addition, we can re-express the 'T HOOFT coupling λ of the gauge theory through string theoretic quantities:

$$\lambda = g_{YM}^2 N_c = g_s N_c = (R/l_s)^4. \qquad (19.14)$$

Here, we inserted the relation $g_{YM}^2 \sim g_s$. Gauge theory computations are perturbative in λ and hence get mapped onto the extremely stringy regime in which the curvature radius R of *AdS₅* is of the order of the string length l_s. Furthermore, the comparison with perturbative string theory results requires the string coupling g_s to be small and hence a large number of colors. It is useful to visualize the relation between gauge and string theory parameters in a 2-dimensional plane; see Figure 19.2. In this picture, the 'T HOOFT coupling increases from left to right while the number N_c of colors grows from bottom to top. Perturbative gauge theory covers a small region on the left side of the diagram. According to the identification of parameters (19.13) and (19.14), the same plane can be parametrized through g_s and R/l_s. The string coupling g_s increases from top to bottom, while the curvature radius R/l_s is very small on the left-hand side and then grows towards the right. Supergravity is valid in the right upper corner of

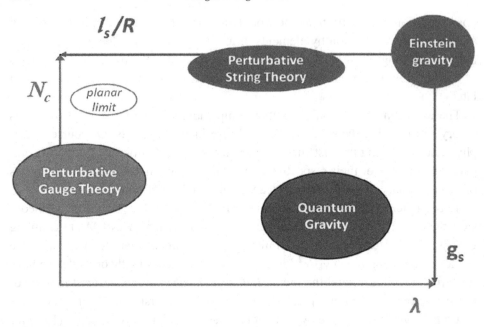

Figure 19.2 The parameter space of gauge/string theory. Gauge theory parameters are shown on the bottom and right side, string theory parameters on the top and left side.

the diagram, far away from the regime where perturbative gauge theory applies. As l_s/R grows, the point particle approximation to string theory becomes invalid, and one needs to resort to genuinely string theoretic tools, such as the ones we have discussed in Chapters 13 to 15. Points on the upper boundary of the diagram actually correspond to the regime of classical (tree-level) string geometry. This regime intersects with the perturbative gauge theory regime in the left upper corner. As we go down in our diagram, we enter the regime of stringy quantum gravity, a regime that is very difficult to access with the existing string theoretic techniques.

19.3 Extending the Holographic Dictionary

Before we discuss tests and applications of the AdS/CFT correspondence, we want to describe a few more entries in the dictionary between the two theories. One of the key observables of a d-dimensional quantum field theory are the connected GREEN's functions. These are typically encoded in a generating functional of the form

$$\mathcal{Z}_\lambda(\Phi_0) := \langle e^{\int d^d x \Phi_0(x)\mathcal{O}(x)} \rangle_\lambda. \tag{19.15}$$

These functionals depend on classical source functions $\Phi_0(x)$, one for each field in the theory. By taking functional derivatives of $\mathcal{Z}$ with respect to the various sources one can recover the connected GREEN's functions from the generating functional $\mathcal{Z}$.

If $\mathcal{O}(x)$ is a scalar field, then the associated source function $\Phi_0(x) = \varphi(x)$ is a scalar as well. For conserved currents $\mathcal{O}_\mu(x) = J_\mu(x)$, on the other hand, the source function should be thought of as a gauge field $\Phi_{0,\mu}(x) = A_\mu(x)$. Indeed the generating functional we introduced above is invariant under gauge transformations $A_\mu \longrightarrow A_\mu + \partial_\mu \Lambda$ since

$$\int d^d x J^\mu(x) \left(A_\mu(x) + \partial_\mu \Lambda(x) \right) = \int d^d x J^\mu(x) A_\mu(x) - \int d^d x \partial_\mu J^\mu \Lambda(x)$$

$$= \int d^d x J^\mu(x) A_\mu(x).$$

Here, we have performed a partial integration and then used the conservation law $\partial_\mu J^\mu = 0$. Similarly, the source for the conserved stress tensor $\mathcal{O}_{\mu\nu} = T_{\mu\nu}$ of a quantum field theory can be considered as a metric $\Phi_{0,\mu\nu} = g_{\mu\nu}$.

In general it is quite a challenge to compute the generating function $\mathcal{Z}_\lambda$ for any higher dimensional interacting gauge theory, in particular for large values of the coupling parameter λ. According to the AdS/CFT correspondence it should be possible to compute $\mathcal{Z}_\lambda$ from string theory. While this is still a daunting task, especially for finite values of the 'T HOOFT coupling, things are expected to simplify when the dual string theory is well approximated by its low-energy supergravity. In the previous section we discussed that the gravity limit in which the string length l_s goes to zero corresponds to the strong coupling limit of the gauge theory. For this regime, Witten proposed the following prescription to compute the generating function [76]:

$$\mathcal{Z}_{\lambda=\infty}(\Phi_0) = e^{S_{\text{SUGRA}}(\Phi^*)}\Big|_{\Phi^*_{\partial AdS}=\Phi_0}. \tag{19.16}$$

On the right-hand side we evaluate the action for type IIB supergravity on $AdS_5 \times S^5$ at a classical solution Φ^*. Since the equation of motion of supergravity are second order, we must prescribe boundary conditions for the fields at $r = \infty$ in order for the solution to be unique. This boundary condition is provided by the source function Φ_0. A precise statement requires us to be a bit more careful about the asymptotic behavior of the classical solutions near the boundary, but this is more of a technical detail for us.

Instead of entering such a discussion let us illustrate the previous formula at one simple example. Our goal is to use eq. (19.16) in order to compute the 2-point function of a scalar field in a d-dimensional field theory. To accomplish this task, we need only a small subsector of the dual supergravity that we assume to take the form

$$S(\Phi) = \frac{1}{2} \int d^{d+1}X \sqrt{-g} \left(g^{\mu\nu} \partial_\mu \Phi \partial_\nu \Phi + m^2 \Phi^2 \right). \tag{19.17}$$

Here $g^{\mu\nu}$ is the metric of AdS_{d+1} that was described in the previous section, at least for $d = 4$. The generalization to other dimensions is obvious. Before we begin our analysis of the action (19.17) we follow usual conventions and introduce a new coordinate $z = r^{-1}$ in which the boundary of AdS now appears at $z = 0$. Inserting the explicit form of the AdS_{d+1} metric, the equations of motion for the free massive scalar field theory (19.17) read

$$\left[\partial_z z^{-d+1} \partial_z + z^{-d+1} \nabla_x^2 - m^2 \right] \Phi = 0. \tag{19.18}$$

After a FOURIER transform in the coordinates x of the boundary, this equation of motion becomes

$$\left[z^2 \partial_z^2 + z \partial_z - (z^2 k^2 + \nu^2) \right] (z^{-d/2} \hat{\Phi}_k(z)) = 0. \tag{19.19}$$

Here, $\hat{\Phi}$ denotes FOURIER transform of the field Φ. The differential equation we have obtained is easily recognized as the modified BESSEL equation with parameter $\nu^2 = d^2/4 + m^2$. It possesses one solution $\hat{\Phi}^*$ that stays regular in the region $z > 0$,

$$z^{-\frac{d}{2}} \hat{\Phi}_k^*(z) \sim K_\nu(\kappa z), \tag{19.20}$$

where K_ν is the modified BESSEL function and $\kappa = |k|$ is the absolute value of the momentum k along the boundary of AdS_{d+1}. A superposition of these solutions with arbitrary coefficients $\hat{\varphi}_0(k)$ gives the most general solution:

$$\Phi^*(z,x) \sim \int d^d k \, e^{ikx} z^{\frac{d}{2}} K_\nu(\kappa z) \kappa^\nu \hat{\varphi}_0(k)$$

$$= \int d^d x' \int d^d k \, e^{ik(x-x')} z^{\frac{d}{2}} K_\nu(\kappa z) \kappa^\nu \varphi_0(x')$$

$$= \int d^d x' \int_0^\infty d\kappa \, J_{d/2-1}(|x - x'|\kappa) \frac{\kappa^{\nu+\frac{d}{2}}}{|x - x'|^{d/2-1}} K_\nu(\kappa z) z^{\frac{d}{2}} \varphi_0(x)$$

$$= \int d^d x' \frac{z^{\nu+\frac{d}{2}}}{(z^2 + |x - x'|^2)^{d/2+\nu}} \varphi_0(x).$$

In this short computation we have used standard properties for a d-dimensional FOURIER transform, including a nice formula that lets us compute the d-dimensional FOURIER transform of a rotationally symmetric function. The latter brought the BESSEL function $J_{d/2-1}$ into the integrand. We then performed the integral over

κ using another well-known integral formula. In this way we obtained a family of solutions of the action (19.17) that depends on the choice of a function $\varphi_0(x)$ on the d-dimensional boundary of AdS_{d+1}.

WITTEN's prescription (19.16) instructs us to insert these solutions back into the original action. The result is

$$S(\Phi^*) = \lim_{\epsilon \to 0} \int_{z=\epsilon} d^d x \sqrt{-g} \Phi^*(\boldsymbol{n} \cdot \nabla)\Phi^* \sim \int d^d x d^d x' \frac{\varphi_0(x)\varphi_0(x')}{|x - x'|^{2(d/2+\nu)}}. \quad (19.21)$$

This step actually requires a bit of care since the naive evaluation of the integral over z is divergent with a divergence arising from the integration region near the boundary at $z = 0$. In order to regularize the expression, the integral is cut off at a finite value $z = \epsilon$. After adding appropriate boundary terms, one can use STOKES's theorem to convert the expression onto a well-defined integral over the boundary at $z = \epsilon$ before sending ϵ to zero; see [76, 62] for more details.

Once we accept the outcome of the previous computation we are finally in a position to calculate the desired 2-point function of a scalar field in d dimensions,

$$\langle \mathcal{O}(x)\mathcal{O}(x') \rangle_{\lambda=\infty} = \frac{\delta}{\delta\varphi_0(x)} \frac{\delta}{\delta\varphi_o(x')} e^{S(\Phi^*)} = \frac{1}{|x - x'|^{2\Delta}}, \quad (19.22)$$

where the weight Δ in the exponent on the right-hand side is given by

$$\Delta = \frac{d}{2} + \sqrt{\frac{d^2}{4} + m^2}. \quad (19.23)$$

The result of our short computation is not that surprising. In fact, what we found is the usual expression for a 2-point function of a scalar field of scaling weight Δ in a d-dimensional scale invariant quantum field theory. On the other hand, in addition to checking that WITTEN's prescription can give a sensible answer, we have also learned how the mass parameter of the scalar field in AdS_5 determines the weight Δ of the associated field in the dual field theory.

Before we close this chapter, we want to phrase some of our findings in a more general context. When dealing with gauge theories, we are interested in (correlators of) gauge invariant fields or operators, such as the stress energy tensor, the trace of the field strength, etc. According to the AdS/CFT dictionary, such operators correspond to the wave functions of closed strings moving in $AdS_5 \times S^5$. Our example of the 2-dimensional scalar field fits into this pattern. The dual massive scalar field in AdS_d could be any of the scalar vibrational modes of the superstring. Similarly, the stress energy tensor of the gauge theory that gets mapped to the massless graviton of the closed string theory, etc. Listing gauge invariant operators in terms of closed string modes may not seem such a big deal at first, until one realizes that this map preserves additional data. In a scale-invariant quantum field theory, operators possess a scaling weight Δ that describes their behavior under re-scalings of space-time. In the case of the scalar field, this scaling weight Δ was related to the mass m of the dual mode by eq. (19.23). While the precise expression for the scaling weight in terms of the mass depends on the nature of the field under consideration, the relation between these two quantities on both sides of the correspondence is very general. The computation of scaling weights in interacting quantum field theories is an important and highly non-trivial problem. The AdS/CFT duality relates it to the determination of the mass spectrum of closed strings in an AdS background. Further exploration of this relation is the main goal in the next chapter.

20

AdS/CFT Correspondence

The aim of this final chapter is to give an idea of the powerful applications of gauge/string dualities at the example of Maldacena's original correspondence between the maximally supersymmetric YANG-MILLS (SYM) theory in four dimensions and the dual string theory on $AdS_5 \times S^5$. We will begin with a brief review of N=4 SYM theory, with special emphasis on its superconformal symmetry. The latter contains a generator of dilations that has been studied extensively after a remarkable observation by Minahan and Zarembo that, in leading order of perturbation theory, the dilation operator of N=4 SYM theory is the Hamiltonian of a HEISENBERG spin chain [44]. We will then turn to the dual string theory and discuss some classical solutions, including that of a folded spinning string found originally by Gubser, Klebanov, and Polyakov in [34]. This solution plays a central role in modern applications of string theory to the multi-color limit of N=4 SYM theory, which we can only just touch upon in a short outlook.

Our discussion will finally bring us back to the very beginning of these chapters and to the origin of string theory as a candidate theory for hadronic physics. It shall discuss how dualities between gauge and string theory manage to circumvent the difficulties physicists faced in the early 1970s in applying string theory to strong interactions.

20.1 N=4 Super YANG-MILLS Theory

We have introduced the field content and action of N=4 SYM theory at the very end of Chapter 11, as a dimensional reduction of 10-dimensional N=1 SYM theory. In addition, we have stated in the previous chapter that N=4 SYM theory gives rise to a conformal quantum field theory. Now we want to explore the model and its conformal symmetry in a bit more detail. Conformal symmetry implies scale invariance, and hence it is a rather unusual property for a 4-dimensional gauge theory. The examples of gauge theories we are usually experiencing in nature, such as, e.g., Quantum Chromodynamics (QCD), are not scale invariant. In fact, at low energies (long distances), QCD is a theory of colorless hadrons, while at very high energies (short distances) its fundamental degrees of freedom are quarks and gluons.

Without much further thought, we would expect N=4 SYM theory to possess N=4 super-POINCARÉ symmetry. The bosonic generators of this superalgebra include the usual generators of a 4-dimensional POINCARÉ algebra, e.g. boosts, rotations and translations. These generate the POINCARÉ algebra $so(1,3) \ltimes \mathbb{R}^{1,3}$. In addition, there are the generators of the R-symmetry $su(4) \sim so(6)$ under which the six scalars of the theory transform as a vector.

In the N=4 POINCARÉ superalgebra, these bosonic generators are supplemented by four multiplets of supercharges, much in the same way as we explained in Chapter 9. As before, we shall denote these supercharges by $Q_{\alpha A}$ and $\tilde{Q}^A_{\dot{\alpha}}$. Here, $A = 1, \ldots, 4$ enumerates the four multiplets. Each of them consists of four components, which we label by $\alpha = 1, 2$ and $\dot{\alpha} = 1, 2$. The relations between all these generators can be spelled out following the general discussion in Chapter 9, and we do not want to repeat this here.

If N=4 SYM theory was a usual relativistic quantum field theory with N=4 supersymmetry, we would be done with the description of its space-time symmetries. On the other hand, we have seen in the previous chapter that the suspected dual string background $AdS_5 \times S^5$ is acted upon by the bosonic algebra $so(2,4) \oplus so(6)$. The latter extends the algebra $so(1,3) \ltimes \mathbb{R}^{1,3} \oplus so(6)$. In fact, while $so(2,4) \cong su(2,2)$ possesses 15 generators, the 4-dimensional POINCARÉ algebra has only 10 of them. The five additional elements are the generators of dilations, denoted D, and of so-called special conformal transformations $K_\mu, \mu = 0, 1, 2, 3$. This additional bosonic symmetry gets further augmented by 16 fermionic generators S^A_α and $\tilde{S}_{\dot{\alpha}A}$. Altogether, the 30 bosonic and 32 fermionic elements span the N=4 superconformal superalgebra $psu(2,2|4)$. For a complete list of all relations and a discussion of further properties; see, e.g., [3].

We conclude that, in order for N=4 SYM theory to be dual to type IIB superstrings in $AdS_5 \times S^5$, the N=4 POINCARÉ supersymmetry of N=4 SYM theory must be augmented to a full superconformal symmetry $psu(2,2|4)$. This is a rather non-trivial feature of the quantum field theory. Here, we want to take a closer look at the scale invariance, which is the most important implication of the symmetry enhancement. Looking back at the action of N=4 SYM theory in Chapter 11 one might be tempted to conclude that scale invariance is obvious. In fact, the classical action is easily seen to be invariant upon re-scaling of all fields $\Phi = A, \phi, \psi$ in the action by a parameter ϖ as

$$\Phi \longrightarrow \Phi_\varpi(x) \equiv \varpi^{-\Delta_{0,\Phi}} \Phi(\varpi^{-1}x),$$

where $\Delta_{0,\Phi}$ is the naive dimension of the field Φ, i.e. $\Delta_{0,A} = 1 = \Delta_{0,\phi}$ and $\Delta_{0,\psi} = 3/2$. But the same is true for all other 4-dimensional gauge theories, including classical massless QCD. In QCD, as well as most other 4-dimensional gauge theories; however, a scale is generated through quantum effects, i.e. the classical scale invariance of the classical model is broken in the quantum theory. It takes a very careful selection of the field content to avoid such a symmetry breaking.

The quantity that signals quantum scale invariance is the so-called beta function of the field theory. It can be computed perturbatively. For a 4-dimensional gauge theory with vector fields (V), scalars (S), and fermions (F) the 1-loop beta function reads [52, 72]:

$$\beta^{\text{1-loop}}(g_{\text{YM}}) \sim g_{\text{YM}}^2 \left(\frac{11}{3} \sum_V C_V - \frac{1}{6} \sum_S C_S - \frac{1}{3} \sum_F C_F \right) + \cdots . \tag{20.1}$$

Here, the sums extend over all fields of a given type, should there be more than one, and C denotes the value of the quadratic CASIMIR element of the color LIE algebra su(N_c) in the corresponding field multiplet. In the case of N=4 SYM theory, we have a single vector field in the adjoint representation of su(N_c). Hence, the first sum consists of a single term, and $C_V = N_c$ is the value of the quadratic CASIMIR element in the adjoint representation. Moving on to the next term, we sum over all the six scalars of N=4 SYM theory. Like the vector field, all scalars of the theory are in the adjoint representation of su(N_c). Hence, the sum in the second term contributes $\sum_S C_S = 6N_c$. Finally, our model contains eight fermions in the adjoint representation giving $\sum_F C_F = 8N_c$. Inserting all these numbers into the expression for the beta function we obtain $\beta = 0$. This demonstrates that quantum N=4 SYM theory is scale invariant, at least to leading (1-loop) order. One may show that the conclusion is not altered by higher order corrections, in complete agreement with the AdS/CFT correspondence.

20.2 Dilation Operator and Spin Chains

As in any conformal field theory, the 2-point functions of scalar primary fields $\mathcal{O}$ of N=4 SYM theory are fixed by conformal symmetry to take the form

$$\langle \mathcal{O}(x)\mathcal{O}(x') \rangle_\lambda = \frac{1}{|x - x'|^{2\Delta(\lambda)}} . \tag{20.2}$$

Here we have explicitly displayed the dependence on the 'T HOOFT coupling λ. It enters the 2-point function only through the scaling weight $\Delta(\lambda) = \Delta_0 + \gamma(\lambda)$. For most fields of N=4 SYM theory, quantum corrections alter the scaling weight compared to its naive (classical) value Δ_0 by a non-vanishing *anomalous dimension*

$\gamma = \gamma(\lambda)$. At weak coupling, i.e. for small values of λ, the anomalous dimension γ is small so that we can approximate the 2-point function through

$$\langle \mathcal{O}(x)\mathcal{O}(x')\rangle_\lambda = \frac{1}{|x - x'|^{2\Delta_0}} \left(1 + \gamma \log |x - x'|^2 + \cdots\right). \qquad (20.3)$$

Hence, to leading order in the coupling, the anomalous dimension may be read off from the coefficient of the logarithmic term in the perturbed 2-point function. We shall make use of this fact below.

Our study of anomalous dimensions in N=4 SYM shall focus on a particular set of fields. In order to introduce them, we consider the following two matrix valued scalar fields:

$$\chi_1 = \frac{1}{\sqrt{2}} \left(\phi_1 + i\phi_2\right), \quad \chi_2 = \frac{1}{\sqrt{2}} \left(\phi_3 + i\phi_4\right).$$

From these we can build gauge-invariant operators of the form

$$\mathcal{O}^L_{a_1,\ldots,a_L} \equiv tr_{N_c} \left(\chi_{a_1} \cdots \chi_{a_L}\right). \qquad (20.4)$$

The multi-index $(a_1, \ldots, a_L)$ parametrizes a set of 2^L fields. Because of the cyclicity of the trace, some of them are identical, though. In fact, two objects $\mathcal{O}^L$ that differ by a cyclic shift of the multi-index are the same, i.e. $\mathcal{O}^L_{a_1,\ldots,a_L} = \mathcal{O}^L_{b_1,\ldots,b_L}$ if there exists an element $\gamma \in \mathbb{Z}_L$ such that $(a_1, \ldots, a_L) = \gamma(b_1, \ldots, b_L)$. Here, γ acts on the multi-index $(b_1, \ldots, b_L)$ by a cyclic shift of the components.

All operators $\mathcal{O}^L$ possess naive scaling weight $\Delta_0 = L$. When we turn on the coupling of N=4 SYM theory, the 2-point function of these fields takes the form

$$\langle \mathcal{O}^L_{\underline{a}}(x)\mathcal{O}^L_{\underline{b}}(x')\rangle_\lambda = \frac{1}{|x - x'|^{2\Delta_0}} \left(I_{\underline{a},\underline{b}} + \Gamma_{\underline{a},\underline{b}} \log |x - x'|^2 + \cdots\right). \qquad (20.5)$$

Here we have introduced the shorthand $\underline{a}$ for the multi-index $\underline{a} = (a_1, \ldots, a_L)$. We can think of the set of all the 2-point functions between our fields as forming a $2^L \times 2^L$ matrix. If we do so, the right-hand side of the previous equation becomes a matrix as well. While the leading term contains the identity matrix I, the coefficient of the logarithmic singularity has a much more complicated matrix structure including off-diagonal elements. This $2^L \times 2^L$-matrix Γ describes the action of a dilation on the objects $\mathcal{O}^L$. Its eigenvectors correspond to special linear combinations of the fields (20.4) with definite scaling weight. The latter is given by the associated eigenvalue of Γ.

At one loop, the matrix Γ was computed by Minahan and Zarembo in the planar limit, i.e. to leading order in the number of colors N_c [44]. The outcome was surprising. What they found was the Hamiltonian of a well-known 1-dimensional spin chain: the Heisenberg magnet:

$$\Gamma \sim \lambda \sum_{l=1}^{L} \left(1 - P_{l,l+1}\right) = \lambda \sum_{l=1}^{L} \left(\frac{1}{2} - 2S_l \cdot S_{l+1}\right). \tag{20.6}$$

Here, we think of the 2^L fields $\mathcal{O}^L$ as elements in the space $(\mathbb{C}^2)^{\otimes L}$ of states of a periodic spin chain of length L. Each site admits two positions of the spin, up and down. These correspond to the two fields χ_1 and χ_2. The lattice sites are enumerated by the index $l = 1, \ldots, L$. We denoted the permutation of two neighboring sites by $P_{l,l+1}$. A simple computation with 4×4 matrices shows that the permutation of neighboring lattice spins can be written in terms of tensor products of Pauli matrices. The latter form the components of the vectors S_l and S_{l+1}. Formula (20.6) for the coefficient of the logarithmic term in the perturbed 2-point function is not that difficult to derive. The permutation arises directly from the structure of the fourth order interaction terms between scalar fields in the action of N=4 SYM theory; see [44] for details.

With the matrix Γ being computed, it remains to determine its eigenvectors and eigenvalues. While this may look like a difficult problem in general, at least two eigenvectors of Γ are very easy to spell out: It is obvious that both $\mathcal{O}_1^L = tr\chi_1^L$ and $\mathcal{O}_2^L = tr\chi_2^L$ are eigenvectors of Γ with eigenvalue $\gamma = 0$. In fact, both fields or rather the associated states of the spin chain are invariant under all permutations $P_{l,l+1}$ for $l = 1, \ldots, L$. Hence they are annihilated by the dilation matrix Γ. In other words, the 1-loop anomalous dimension of both these operators vanishes. It turns out the statement continues to hold at higher loops. Actually, the vanishing of the anomalous dimension for $\mathcal{O}_1^L$ and $\mathcal{O}_2^L$ is a rather direct consequence of supersymmetry since the two operators are chiral primaries.

The other eigenvalues of the matrix Γ are not quite as easy to find. But there exists some beautiful technology that was developed in statistical physics in order to solve the HEISENBERG spin chain. The eigenstates of the HEISENBERG Hamiltonian describe magnons that propagate with some rapidity θ along the chain. The energy of these magnons is given by some dispersion law $E = E(\theta)$. Multiple magnon states give rise to eigenstates, provided the set of rapidities θ_i satisfy the so-called BETHE Ansatz equations [10]. The associated eigenvalues are obtained by summing the energies $E(\theta_i)$ of the individual magnons; see e.g. [22].

This concludes our brief description of 1-loop anomalous dimensions in N=4 SYM theory. Let us just add that the expression (20.6) for Γ, and/or the BETHE Ansatz equations that were derived from it, have been generalized in several different directions. These include considering more general operators, higher orders, and also gauge theories with less supersymmetry; see [8] for a review.

20.3 Strings in the $AdS_5 \times S^5$ Background

We now turn to the dual theory of strings in an AdS_5 background. Going into a construction of this curved string background is well beyond the scope of our text; see, however, [8] for a status review. Instead we shall focus on the semiclassical approximation, which can be trusted at least for large quantum numbers. A much more comprehensive discussion of the following material can be found e.g. in [69].

Classical Equations of Motion

If we think of the 5-dimensional factors AdS_5 and S^5 as surfaces in a 6-dimensional embedding space, the action for bosonic fields in the string theoretic description of $AdS_5 \times S^5$ takes the form

$$S[Y,X] = -\frac{R^2}{4\pi\alpha'} \int d\tau d\sigma$$

$$\left[\partial_a Y_P \partial^a Y^P + \tilde{\Lambda}(Y_P Y^P + 1) + \partial_a X_M \partial^a X^M + \Lambda(X_M X^M - 1) \right]. \quad (20.7)$$

Here, we have promoted the six coordinates of the embedding space for AdS_5 to fields Y^P on the world-surface. The indices $P = -1, 0, 1, \ldots, 3$ are raised on lowered with the metric of $\mathbb{R}^{2,4}$; see the previous chapter. The defining equation (19.9) of the embedded AdS_5 is implemented with the help of a LAGRANGE multiplier $\tilde{\Lambda}$. Similar comments apply to the description of the sphere S^5 through the six fields X^M with $M = 1, \ldots, 6$. In this case, indices are raised and lowered with the usual EUCLIDEAN metric.

From the action we can infer the following simple equations of motion:

$$\partial_a \partial^a Y^P - \tilde{\Lambda} Y^P = 0, \quad Y_P Y^P = -1, \quad (20.8)$$

$$\partial_a \partial^a X^M + \Lambda X^M = 0, \quad X_M X^M = 1. \quad (20.9)$$

As usual in string theory, these equations of motion are to be supplemented by VIRASORO constraints. From the condition $T_{00} = T_{11} = 0$ we obtain

$$\dot{Y}_P \dot{Y}^P + Y'_P Y'^P + \dot{X}_M \dot{X}^M + X'_M X'^M = 0. \quad (20.10)$$

Similarly, from $T_{01} = 0 = T_{10}$ one finds

$$\dot{Y}_P Y'^P + \dot{X}_M X'^M = 0. \quad (20.11)$$

To complete our short review of the classical theory, we want to spell out expressions for the bosonic generators of the superconformal symmetry algebra. These read

$$M_{PQ} = \frac{R^2}{2\pi\alpha'} \int d\sigma \left(Y_P \dot{Y}_Q - Y_Q \dot{Y}_P \right), \tag{20.12}$$

$$R_{MN} = \frac{R^2}{2\pi\alpha'} \int d\sigma \left(X_M \dot{X}_N - X_N \dot{X}_M \right). \tag{20.13}$$

The first set of equations define the generators of the conformal algebra so(2, 4), while the expressions for R_{MN} give the generators of the R-symmetry so(6).

Classical Solutions: Point-Like Strings

The first set of classical solutions we want to discuss describe point-like strings that rotate on a circle inside the 5-sphere S^5 of the string background $AdS_5 \times S^5$. These solutions are given by

$$Y_{-1} + iY_0 = e^{i\kappa\tau}, \quad Y_P = 0, P = 1, \ldots, 4, \tag{20.14}$$

$$X_1 + iX_2 = e^{i\omega\tau}, \quad X_M = 0, M = 3, \ldots, 6. \tag{20.15}$$

It is straightforward to check that the functions Y_P and X^M satisfy the equations of motion (20.8, 20.9) with the LAGRANGE multipliers $\tilde{\Lambda} = -\kappa^2$ and $\Lambda = \omega^2$. They also solve the VIRASORO constraints, provided we choose $\omega^2 = \kappa^2$.

On these solutions, the generator of dilations $D = M_{-1,0}$ and the generator of the R-charge $J_1 = R_{12}$ assume the values

$$D \sim \frac{\sqrt{\lambda}}{2\pi}\kappa, \quad J_1 \sim \frac{\sqrt{\lambda}}{2\pi}\kappa. \tag{20.16}$$

We observe that $D = R_{12}$. This relation between the conformal weight and the R-charge $J = R_{12}$ suggests a direct relation with the operators $\mathcal{O}_1^L$ of N=4 SYM theory, which we discussed in the previous section. Note that the naive conformal weight of these fields is $\Delta_0 = L$. As we observed in our discussion after eq. (20.6), the conformal weight of $\mathcal{O}_1^L$ does not receive a 1-loop correction. In fact, it does not receive corrections at any loop order, i.e., $\Delta(\lambda) = \Delta_0 = L$. On the other hand, the constituent field χ_1 from which we build $\mathcal{O}_1^L$ is easily seen to be an eigenstate of R_{12} with eigenvalue $Q = 1$. Consequently, $\mathcal{O}_1^L$ carries the R-charge $Q = L$. Since R_{12} is a compact generator, its spectrum is discrete so that it cannot depend on the 'T HOOFT coupling λ. Hence the operators $\mathcal{O}_1^L$ satisfy $D = \Delta_0 = L = Q = R_{12}$ for any value of λ. Under the AdS/CFT duality, these fields correspond to the point-like string solutions we constructed above.

Classical Solutions: Folded Spinning String

The second solution we want to discuss describes a folded string that spins in an AdS_3 subspace of AdS_5. The solution was first discussed by Gubser, Klebanov, and Polyakov in [34]. These authors suggested the following Ansatz:

$$Y_{-1} + iY_0 = \cosh \rho(\sigma) e^{i\kappa\tau}, \tag{20.17}$$

$$Y_1 + iY_2 = \sinh \rho(\sigma) e^{i\omega\tau} \tag{20.18}$$

with all other fields set to zero. From the VIRASORO constraints one can easily obtain that

$$\rho' = \partial_\sigma \rho, \quad (\rho')^2 = \kappa^2 \cosh^2 \rho - \omega^2 \sinh^2 \rho. \tag{20.19}$$

We can use the second equation to deduce the following expression for the line element $d\sigma$ of the world-sheet coordinate σ:

$$d\sigma = \frac{d\rho}{\sqrt{\kappa^2 \cosh^2 \rho - \omega^2 \sinh^2 \rho}}. \tag{20.20}$$

Let us observe that $d\sigma$ becomes singular for $\rho = \rho_0 = \omega/\kappa$. This value represents the turning point of the folded string in AdS_3. We actually do not have to construct the explicit solution in order to compute the associated value of the dilation generator $D = M_{-1,0}$ and the spin $S = M_{12}$. These take the following form:

$$D \sim \frac{\sqrt{\lambda}}{2\pi} \int_0^{2\pi} d\sigma \kappa \cosh^2 \rho \sim \sqrt{\lambda} \int_0^{\rho_0} d\rho \frac{\kappa \cosh^2 \rho}{\sqrt{\kappa^2 \cosh^2 \rho - \omega^2 \sinh^2 \rho}}, \tag{20.21}$$

$$S \sim \frac{\sqrt{\lambda}}{2\pi} \int_0^{2\pi} d\sigma \omega \sinh^2 \rho \sim \sqrt{\lambda} \int_0^{\rho_0} d\rho \frac{\omega \sinh^2 \rho}{\sqrt{\kappa^2 \cosh^2 \rho - \omega^2 \sinh^2 \rho}}. \tag{20.22}$$

We want to analyze this result for D and S in the long string limit in which the turning point ρ_0 of the folded string extends all the way to the boundary of the AdS space, i.e., $\rho_0 \to \infty$. In this limit we write

$$\coth \rho_0 = \frac{\omega}{\kappa} = 1 + 2\eta$$

with some small positive parameter η. We can then expand ρ_0 as $2\rho_0 \sim \ln(1/\eta) + \cdots$. Upon insertion into the integral formulas for D and S one obtains

$$D \sim \frac{\sqrt{\lambda}}{2\pi} \left(\frac{1}{\eta} + \ln \frac{1}{\eta} + \cdots \right), \tag{20.23}$$

$$S \sim \frac{\sqrt{\lambda}}{2\pi} \left(\frac{1}{\eta} - \ln \frac{1}{\eta} + \cdots \right). \tag{20.24}$$

If we denote the values of D and S by Δ and S, respectively, we finally conclude

$$\Delta - S = \frac{\sqrt{\lambda}}{\pi} \ln \frac{1}{\eta} + \cdots = \frac{\sqrt{\lambda}}{\pi} \ln \frac{S}{\sqrt{\lambda}} + \cdots . \tag{20.25}$$

Historically this was a very important result of [34]. Unlike the previous example of point-like strings, which described protected operators whose scaling weight $\Delta(\lambda) = \Delta_0$ does not depend on the coupling, the scaling weight of operators dual to the spinning folded string depends on the 'T HOOFT coupling λ.

The relevant fields of N=4 SYM theory are not that difficult to guess. Through a comparison of quantum numbers, similarly to our discussion for point-like string solutions, one is led to

$$\mathcal{O}_S \sim tr(\bar{\chi}_1 D_+^S \chi_1).$$

Note that the classical string analysis of Gubser et al. can be trusted only in the limit of large quantum numbers, i.e., in the limit of large spin S. In this limit it suggests that

$$\Delta(\lambda) - S = f(\lambda) \ln S + \cdots .$$

where the dots stand for lower order terms in the spin S. The coefficient $f(\lambda)$ is known as the universal cusp anomalous dimension. Our discussion of folded spinning strings above gives $f(\lambda) \sim \sqrt{(\lambda)}/\pi$ for the leading term at strong 'T HOOFT coupling. This analysis can be extended to include subleading contributions:

$$f(\lambda) = \frac{\sqrt{\lambda}}{\pi} - \frac{3 \ln 2}{\pi} + \cdots \qquad \text{for} \quad \lambda \gg 1.$$

Similarly, perturbative gauge theory computations provided the first few terms at weak coupling:

$$f(\lambda) = \frac{1}{2} \frac{\lambda}{\pi^2} - \frac{1}{96\pi^2} \lambda^2 + \cdots \qquad \text{for} \quad \lambda \ll 1.$$

Of course, these expansions at weak and strong coupling call for an interpolation. This was found in the seminal paper [7], where the authors proposed an integral equation, now known as the BEISERT-EDEN-STAUDACHER (BES) equation, that allows us to determine the cusp anomalous dimension $f(\lambda)$ all the way from weak to strong coupling. It reproduces all known terms from perturbative gauge theory near $\lambda \sim 0$ and from classical string theory at $\lambda = \infty$. This was actually the first time the anomalous dimension of an unprotected operator was computed to all orders in the 'T HOOFT coupling. It set off an extended research program to determine the anomalous dimensions of all local operators in N=4 SYM theory. The task has been completed over the last few years; see [32] for a recent status report and references. In gauge theory, these results on anomalous dimensions represent a breakthrough that was possible only through the correspondence with string theory in AdS backgrounds.

Strings and High Energy Scattering

Throughout the last few sections we began to see some of the intriguing new developments in gauge theory that resulted from the duality with string theory. In fact, string theory in an AdS background has provided all-loop results of quantities that are very difficult to compute with traditional gauge theory methods even at two or three loops. This nurtures hopes that string theory could uncover a radically new way of thinking about gauge theory. Of course, at the moment this progress is still restricted to some special gauge theories and to the planar limit.

All these remarkable new insights into gauge theory should be contrasted with our discussion in the very first chapter where we discussed the early unsuccessful attempts to apply string theory to hadronic physics. Let us briefly review a few elements from that discussion. There, we looked at a $2 \to 2$ scattering process with kinematic invariants s and t. In flat space string theory, such amplitudes are well known. If we consider, for example, the scattering of four closed string tachyons, the answer is given by eq. (4.29). Keeping the MANDELSTAM variable t fixed, we can evaluate the large s behavior with the help of STIRLING's formula for Γ functions. The result is

$$\mathcal{A}^{2 \to 2}(s, t) \longrightarrow s^{2 + \frac{\alpha'}{2} t}. \tag{20.26}$$

In the so-called REGGE regime $t > 0$, the amplitude has very desired properties. In fact, when $t = M_J^2$ is chosen such that $2 + \alpha' M_J^2 / 2 = J$ is integer, then the amplitude $\mathcal{A}(s, M_J^2) \sim s^J$ describes the exchange of a particle with spin J. Such a linear relation between masses M_J^2 and spins J has indeed been experimentally observed for meson resonances, as we have discussed. In the physical regime with $t = -\frac{s}{4}(1 - \cos \theta_S) < 0$, the behavior is very different. For sufficiently large $|t|$, the amplitude falls off as $\mathcal{A} \sim s^{-a}$ and hence much faster than it is observed experimentally. The rapid fall-off is a consequence of the string's extended nature. It was this observation that made us abandon string theory as a promising candidate for a model of hadronic physics.

But in our description of the AdS/CFT correspondence we have concluded that strings can describe particle theories, potentially even more successfully than gauge theories. How is this compatible? In order to resolve the apparent tension, one may consider [53] a caricature of the way string theory computed scattering amplitudes in gauge theory:

$$\mathcal{A}_{\text{GT}}(s, t) \sim \int dr \prod_{i=1}^{4} \psi_i(r) \mathcal{A}_{\text{ST}}(\tilde{s}, \tilde{t}). \tag{20.27}$$

On the gauge theory side, we study the scattering of some objects that are described by gauge invariant operators. According to the dictionary of the AdS/CFT correspondence, these may be represented as some particular wave functions of the closed bosonic string in AdS_5. We denoted these wave functions by ψ_i and emphasized their dependence on the radial coordinate. The function $\psi_i(r)$ provides the probability to find a closed string at radius r. Each gauge invariant operator comes with its own characteristic probability distribution. Closed strings at radius r scatter with the amplitude $\mathcal{A}_{ST}$. Some care must be taken in determining the kinematic invariants of the latter. In order to see how these are related to the kinematic invariants of the gauge theory process, we recall that the motion of strings in a RIEMANNIAN manifold is described by

$$S[X] \sim \frac{1}{2\pi\alpha'} \int d^2z \, g_{\mu\nu} \partial X^\mu \bar{\partial} X^\nu + \cdots = \frac{1}{2\pi\alpha'} \int d^2z \, \frac{r^2}{R^2} \partial X^\mu \bar{\partial} X_\mu.$$

Here, we have inserted the line element $ds^2 = dx^2 r^2/R^2$ for the directions x_0, x_1, x_2, x_3 of AdS_5. Of course, the letter r denotes one of the fields of our worldsheet model, and the corresponding theory is highly non-linear and therefore difficult to construct. But we can gain some qualitative understanding by observing that the above action looks as if we would study strings whose effective α'_{eff} depends on the radius r through

$$\alpha'_{\text{eff}}(r) = \alpha' R^2/r^2. \tag{20.28}$$

In other words, strings at large values r have a very large tension and hence appear essentially point-like. For small values of r, on the other hand, the string tension goes to zero so that very little energy is needed to excite closed string vibrations. This behavior of the effective string tension can be attributed to the red-shift in the vicinity of the D3-brane. Coming back to our scattering amplitude we conclude that

$$\mathcal{A}_{GT}(s,t) \sim \int dr \prod_{i=1}^{4} \psi_i(r) \, \tilde{s}^{-2 + \frac{\alpha' R^2}{2r^2} t}. \tag{20.29}$$

Here, $\tilde{s}$ is the red-shifted energy $\tilde{s} \sim Rs/r$ of the process. Now let us see what this formula implies on the behavior of the scattering amplitude. In the REGGE regime with $t > 0$, the amplitude is dominated by the contributions arising from scattering processes near $r \sim 0$. In this region of AdS_5, the closed strings possess a small effective string tension so that their excitations might give rise to the observed meson resonances at low energies. When $t < 0$, on the other hand, the dominant contribution to the amplitude comes from scattering of strings near $r \sim \infty$. In this region, the strings have extremely large tension so that they behave like point-like objects. Therefore, string scattering in AdS geometries is guaranteed to produce the hard, particle-like, cross sections that are observed in nature.

There is another important lesson we learn from here. The high-energy (short distance) features of the gauge theory (such as asymptotic freedom, etc.) are related to the region at $r = \infty$. Low-energy (long distance) properties (such as confinement, meson resonances, etc.), on the other are encoded in the geometry near $r \sim 0$. In the case of AdS_5, there are no interesting long distance features. Note that the effective string length (20.28) diverges at $r = 0$. This means that all the scattering amplitudes in the REGGE regime are dominated by contributions from $r = 0$ where it requires infinitesimally small amounts of energy to excite string oscillations. In other words, the REGGE trajectory has infinite slope. In realistic models of confinement, the effective string length must be bounded from above. This feature as well as many other properties that are desirable in order to describe physically relevant models can indeed be implemented in gauge/gravity dualities; see, e.g., [48, 1] for an overview and references to the vast literature on the subject.

In conclusion, for a string theory description of gauge theory to work, it is absolutely crucial that the strings propagate an a 5-dimensional curved geometry. Through the r-dependent red-shift factor string theory can reproduce both hard high-energy scattering and low-energy meson resonances. The early attempts to describe gauge theory physics through strings in 4-dimensional flat space (times a compactification space) did not incorporate this crucial feature and hence were bound to fail. But this failure was due to a missing insight that is overcome by the AdS/CFT correspondence, opening the way to intriguing and powerful new applications of strings to quantum field theory. In fact, for planar N=4 SYM theory, string theory has even delivered exact formulas that describe the high-energy limit of scattering amplitudes with multiple external gluons, at both infinite [4] and intermediate [5] gauge theory coupling. This demonstrates the ability of string theory to provide new insights into gauge theory even in the arena of high-energy scattering where strings had once taken such a battering. What may have seemed like wishful thinking in the introduction when we went through our sequence of suggestive images has finally obtained very concrete contours.

Bibliography

[1] M. Ammon and J. Erdmenger. *Gauge/gravity Duality*. Cambridge University Press, 2015.

[2] G. Arutyunov. *Lectures on String Theory*. 2009. Utrecht University.

[3] G. Arutyunov and S. Frolov. Foundations of the $AdS_5 \times S^5$ superstring. *Part I. J. Phys.*, A42:254003, 2009.

[4] J. Bartels, V. Schomerus, and M. Sprenger. The Bethe roots of Regge cuts in strongly coupled $\mathcal{N} = 4$ SYM theory. *JHEP*, 07:098, 2015.

[5] B. Basso, S. Caron-Huot, and A. Sever. Adjoint BFKL at finite coupling: a short-cut from the collinear limit. *JHEP*, 01:027, 2015.

[6] K. Becker, M. Becker, and J. H. Schwarz. *String Theory and M-Theory: A Modern Introduction*. Cambridge University Press, 2006.

[7] N. Beisert, B. Eden, and M. Staudacher. Transcendentality and crossing. *J. Stat. Mech.*, 0701:P01021, 2007.

[8] N. Beisert et al. Review of AdS/CFT integrability: an overview. *Lett. Math. Phys.*, 99:3–32, 2012.

[9] A. A. Belavin, A. M. Polyakov, and A. B. Zamolodchikov. Infinite conformal symmetry in two-dimensional quantum field theory. *Nucl. Phys.*, B241:333–380, 1984.

[10] H. Bethe. On the theory of metals. 1. Eigenvalues and eigenfunctions for the linear atomic chain. *Z. Phys.*, 71:205–226, 1931.

[11] V. Bouchard. *Lectures on Complex Geometry, Calabi-Yau Manifolds and Toric Geometry*. 2007.

[12] L. Brink, P. Di Vecchia, and P. S. Howe. A locally supersymmetric and reparametrization invariant action for the spinning string. *Phys. Lett.*, B65:471–474, 1976.

[13] R. C. Brower and K. A. Friedman. Spectrum generating algebra and no ghost theorem for the Neveu-Schwarz model. *Phys. Rev.*, D7:535–539, 1973.

[14] R. C. Brower. Spectrum generating algebra and no ghost theorem for the dual model. *Phys. Rev.*, D6:1655–1662, 1972.

[15] C.-S. Chu and P.-M. Ho. Noncommutative open string and D-brane. *Nucl. Phys.*, B550:151–168, 1999.

[16] P. D. B. Collins. *An Introduction to Regge Theory and High-Energy Physics*. Cambridge Monographs on Mathematical Physics. Cambridge University Press, 2009.

[17] S. Deser and B. Zumino. A complete action for the spinning string. *Phys. Lett.*, B65:369–373, 1976.

[18] P. Di Francesco, P. Mathieu, and D. Senechal. *Conformal Field Theory*. Springer Verlag, 1997.

[19] R. Dijkgraaf, C. Vafa, E. P. Verlinde, and H. L. Verlinde. The operator algebra of orbifold models. *Commun. Math. Phys.*, 123:485, 1989.

[20] J. Dixmier. *Von Neumann Algebras*. North-Holland Mathematical Library. North-Holland, 1981.

[21] M. J. Duff, R. R. Khuri, and J. X. Lu. String solitons. *Phys. Rept.*, 259:213–326, 1995.

[22] L. D. Faddeev. How algebraic Bethe ansatz works for integrable model. In *Relativistic Gravitation and Gravitational Radiation. Proceedings, School of Physics, Les Houches, France, September 26–October 6, 1995*, pp. 149–219, 1996.

[23] D. Gepner. Space-time supersymmetry in compactified string theory and superconformal models. *Nucl. Phys.*, B296:757, 1988.

[24] F. Gliozzi, J. Scherk, and D. I. Olive. Supergravity and the spinor dual model. *Phys. Lett.*, B65:282, 1976.

[25] P. Goddard, J. Goldstone, C. Rebbi, and C. B. Thorn. Quantum dynamics of a massless relativistic string. *Nucl. Phys.*, B56:109–135, 1973.

[26] P. Goddard and C. B. Thorn. Compatibility of the dual pomeron with unitarity and the absence of ghosts in the dual resonance model. *Phys. Lett.*, B40:235–238, 1972.

[27] M. B. Green, J. H. Schwarz, and E. Witten. *Superstring Theory. Vol. 1: Introduction*. Cambridge University Press, 1987.

[28] M. B. Green, J. H. Schwarz, and E. Witten. *Superstring Theory. Vol. 2: Loop Amplitudes, Anomalies and Phenomenology*. Cambridge Univetsity Press, 1987.

[29] M. B. Green and J. H. Schwarz. Anomaly cancellation in supersymmetric D=10 gauge theory and superstring theory. *Phys. Lett.*, B149:117–122, 1984.

[30] B. R. Greene. *Lectures on the Quantum Geometry of String Theory*. In *Quantum Symmetries*, Proceedings of the NATO Advanced Study Institute, 64th Session, Les Houches, France, August 1–September 8, 1995, edited by A. Connes, K. Gawedzki, and J. Zinn-Justin, 126–193, Elsevier, 1995.

[31] P. A. Griffiths and J. Harris. *Principles of Algebraic Geometry*. Wiley, 1978.

[32] N. Gromov, F. Levkovich-Maslyuk, and G. Sizov. Quantum Spectral Curve and the Numerical Solution of the Spectral Problem in AdS5/CFT4. 2015.

[33] D. J. Gross, J. A. Harvey, E. J. Martinec, and R. Rohm. The heterotic string. *Phys. Rev. Lett.*, 54:502–505, 1985.

[34] S. S. Gubser, I. R. Klebanov, and A. M. Polyakov. A semiclassical limit of the gauge/string correspondence. *Nucl. Phys.*, B636:99–114, 2002.

[35] M. Henneaux and C. Teitelboim. *Quantization of Gauge Systems*. Princeton Series in Physics. Princeton University Press, 1992.

[36] J. Hoppe. *Membranes and Matrix Models*. 2002. https://arxiv.org/abs/hep-th/0206192.

[37] A. C. Irving and R. P. Worden. Regge phenomenology. *Phys. Rept.*, 34:117–231, 1977.

[38] V. G. Kac. *Infinite Dimensional Lie Algebras*. Cambridge University Press, 1990.

[39] E. Kiritsis. *String Theory in a Nutshell*. Princeton University Press, 2007.

[40] D. Kutasov. Introduction to little string theory. In *Superstrings and Related Matters. Proceedings, Spring School, Trieste, Italy, April 2–10, 2001*, pp. 165–209, 2001.

[41] D. Lüst and S. Theisen. *Lectures on String Theory*. Lecture Notes in Physics. Springer Verlag, 1989.

[42] J. M. Maldacena. The large N limit of superconformal field theories and supergravity. *Adv. Theor. Math. Phys.*, 2:231–252, 1998.

[43] J. E. Marsden and T. Ratiu. *Introduction to Mechanics and Symmetry: A Basic Exposition of Classical Mechanical Systems*. Texts in Applied Mathematics. Springer, 1994.

[44] J. A. Minahan and K. Zarembo. The Bethe ansatz for N=4 superYang-Mills. *JHEP*, 03:013, 2003.

[45] C. Montonen and D. I. Olive. Magnetic monopoles as gauge particles? *Phys. Lett.*, B72:117, 1977.

[46] D. Mumford. *Tata Lectures on Theta, 1*. Modern Birkhuser Classics. Springer, 2007.

[47] Y. Nambu. *Dual model of hadrons*. 1970. University of Chicago Preprint EFI-70-07.

[48] H. Nastase. *Introduction to the AdS/CFT Correspondence*. Cambridge University Press, 2015.

[49] A. Neveu and J. Scherk. Connection between Yang-Mills fields and dual models. *Nucl. Phys.*, B36:155–161, 1972.

[50] H. B. Nielsen and P. Olesen. A parton view on dual amplitudes. *Phys. Lett.*, B32:203, 1970.

[51] J. E. Paton and H.-M. Chan. Generalized Veneziano model with isospin. *Nucl. Phys.*, B10:516–520, 1969.

[52] M. E. Peskin and D. V. Schroeder. *An Introduction to Quantum Field Theory*. Westview Press, 1995.

[53] J. Polchinski and M. J. Strassler. Hard scattering and gauge/string duality. *Phys. Rev. Lett.*, 88:031601, 2002.

[54] J. G. Polchinski. *Lectures on D-branes*. 1996. https://arxiv.org/abs/hep-th/9611050.

[55] J. G. Polchinski. *String Theory – Volume I*. Cambridge Monographs on Mathematical Physics. Cambridge University Press, 2005.

[56] J. G. Polchinski. *String Theory – Volume II*. Cambridge Monographs on Mathematical Physics. Cambridge University Press, 2005.

[57] A. Recknagel and V. Schomerus. *Boundary Conformal Field Theory and the Worldsheet Approach to D-branes*. Cambridge University Press, 2013.

[58] J. Scherk and J. H. Schwarz. Dual models for nonhadrons. *Nucl. Phys.*, B81:118–144, 1974.

[59] V. Schomerus. D-branes and deformation quantization. *JHEP*, 9906:030, 1999.

[60] N. Seiberg and E. Witten. String theory and noncommutative geometry. *JHEP*, 9909:032, 1999.

[61] S. Sethi and M. Stern. D-brane bound states redux. *Commun. Math. Phys.*, 194:675–705, 1998.

[62] K. Skenderis. Lecture notes on holographic renormalization. *Class. Quant. Grav.*, 19:5849–5876, 2002.

[63] R. Slansky. Group theory for unified model building. *Phys. Rept.*, 79:1–128, 1981.

[64] L. Susskind. Dual symmetric theory of hadrons. 1. *Nuovo Cim.*, A69:457–496, 1970.

[65] R. J. Szabo. *An Introduction to String Theory and D-brane Dynamics*. Imperial College Press, 2004.

[66] G. 't Hooft. A planar diagram theory for strong interactions. *Nucl. Phys.*, B72:461, 1974.

[67] J. Terning. *Modern Supersymmetry: Dynamics and Duality*. Oxford University Press, 2006.

[68] D. Tong. *Lectures on String Theory*. 2009. http://www.damtp.cam.ac.uk/user/tong/string.html.

[69] A. A. Tseytlin. Review of AdS/CFT integrability, chapter II.1: classical $AdS_5 \times S^5$ string solutions. *Lett. Math. Phys.*, 99:103–125, 2012.

[70] A. van Proeyen and D. Z. Freedman. *Supergravity*. Cambridge University Press, 2012.

[71] G. Veneziano. Construction of a crossing-symmetric, Regge behaved amplitude for linearly rising trajectories. *Nuovo Cim.*, A57:190–197, 1968.

[72] S. Weinberg. *The Quantum Theory of Fields. Vol. 1: Foundations.* Cambridge University Press, 1995.

[73] J. Wess and J. Bagger. *Supersymmetry and Supergravity.* Princeton Series in Physics. Princeton University Press, 1992.

[74] P. West. *Introduction to Strings and Branes.* Cambridge University Press, 2012.

[75] P. C. West. *Introduction to Supersymmetry and Supergravity.* World Scientific, 1986.

[76] E. Witten. Anti-de Sitter space and holography. *Adv. Theor. Math. Phys.*, 2:253–291, 1998.

[77] T. Yoneya. Quantum gravity and the zero slope limit of the generalized Virasoro model. *Lett. Nuovo Cim.*, 8:951–955, 1973.

[78] T. Yoneya. Connection of dual models to electrodynamics and gravidynamics. *Prog. Theor. Phys.*, 51:1907–1920, 1974.

[79] B. Zumino. Supersymmetry and Kahler mainfolds. *Phys. Lett.*, B87:203, 1979.

[80] B. Zwiebach. *A First Course in String Theory.* Cambridge University Press, 2005.

Index

Printed in the United States
By Bookmasters